SCIENCE OR FICTION?

The Phony Side of Particle Physics

OFER COMAY

SCIENCE OR FICTION?

The Phony Side of Particle Physics

OFER COMAY

Scientific Advisor
Eliyahu Comay

SAMUEL WACHTMAN'S SONS   DEKEL ACADEMIC PRESS

SCIENCE OR FICTION?

The Phony Side of Particle Physics

OFER COMAY
Copyright © 2014

Dekel Academic Press
www.dekelpublishing.com
North American rights by
Samuel Wachtman's Sons, Inc.
ISBN 978-1-888820-81-2

Editor:	Zvi Morik
Language editing:	Kathleen Roman
Scientific Advisor:	Eliyahu Comay

Cover image: Atom © Roywylam / Dreamstime.com

Cover design and typesetting by

DESIGN PEAKS®

For information contact:

Dekel Publishing House	**Samuel Wachtman's Sons, Inc.**
P.O. Box 45094	2460 Garden Road, Suite C
Tel Aviv 6145002, Israel	Monterey, CA 93940, U.S.A.
Tel: +972 3506-3235	Tel: 831 649-0669
Fax: +972 3506-7332	Fax: 831 649-8007
Email: info@dekelpublishing.com	Email: samuelwachtman@gmail.com

TABLE OF CONTENTS

_____

5

THE CAVE

Acknowledgements

I wish to thank my family, and in particular Oded, Dror, and Omri for their many contributing comments, my friend Paz Einat who helped me find the latest relevant publications, and my friend Yoram Tal, who helped me make this book simpler and more accessible.

I wish to thank the students who read my book and provided me with notes and opinions, and who were kind enough to mention both the parts that made them laugh out loud *and* those that bored them silly.

And finally, I wish to thank the physics professors who gave many hours of their time for this project, but who nevertheless preferred to stay anonymous.

Foreword

This book had to be rewritten several times before achieving its current form. The story it tells is nothing short of incredible and pertains to the forefront of scientific thought. Therefore, it contains several issues that required my special attention.

I cannot overstate the significance of reliability and objectivity for anyone who argues that a leading scientific theory is, in fact, unfounded. For this reason, special care has been taken to verify that every single scientific datum included in this book has been published in reputable scientific journals. The data prove that contradictions to leading scientific theories have been published by genuine, unbiased scientists. The book describes a large number of events wherein scientists had stumbled upon an observation that contradicted a leading theory. These scientists, however, were unaware of how prevalent such discrepancies in fact were, as, rather strangely, data of this kind are never mentioned in textbooks. This book examines a comprehensive list of such contradictions for the first time, and seeks to shed light on their internal logic and significance.

In order to allow readers to assess the book's impartiality on their own, the book comes complete with references and excerpts whenever deemed necessary. Consequently, a substantial portion of this book takes the form of references and citations.

Another issue forced upon me was the vitriolic campaign of delegitimization against any scientist who dares to question or cast doubts on existing theories. I hope I have managed to treat the matter gently, albeit with the firmness and resolve I believe it merits. Nevertheless, I would like to preemptively apologize if from time to time I was unable to help myself from hurling the occasional sarcastic remark at those who partake in that campaign.

My third stumbling block stems from the twofold purpose this book seeks to attain. The object of this book is not only to present a satisfactory account of discrepancies found in current scientific theories, but also to provide popular science readers with tools that will enable them to fully understand the controversy, and to judge the significance and likelihood of each claim for themselves. The scientific contents of this book are usually deemed obscure and unpresentable to popular science readers. Nevertheless, I hope that my efforts were sufficient in rendering them accessible to everyone. To that end, I persevered in addressing only those scientific concepts that are necessary for understanding the arguments presented in this book. I used a great deal of illustrations, some of which, I admit, contain a simplified version of reality. That said, the scientific basis of this book is sound, and I have endeavored to make it explicitly clear whenever an approximation of reality is presented to readers.

Several readers mentioned that the book is challenging because it forces them to think and take a stand. In one chapter this approach is used to extremes and readers are asked to solve scientific problems that even the current theory cannot solve. I hope that readers will find this challenge rewarding.

The Chronological Evidence Test

Popular science books habitually present scientific theories based on the chronological order in which the evidence was found. This method can be useful when seeking to explain *why scientists adhere to a certain theory*. In this book, however, discoveries are often presented in a non-chronological fashion. The reason for this is that this book has no intention of examining *why* a certain theory is preferable to others. It is more interesting to try and find out which theory *is actually correct*. At times, the non-chronological portrayal of events might make it seem as though I am trying to ridicule the scientists who adhere to certain theories, and I must stress that this was never my intent. I know it is very difficult to abandon a concept that seemed reasonable at the time, even as new evidence comes to light.

But one of the most important tests, I believe, of a scientific theory's veracity, is the chronological evidence test. Such a test requires a great deal of scientific integrity and forces the scientist to answer the following question: Had the scientific findings we currently possess first been discovered in a different sequence, would we have also adhered to the theories we now consider to be true? *Correct* theories never depend on the order in which the evidence is discovered. In this book, I seek to demonstrate that the theory adhered to by most contemporary scientists today appears completely unfounded when the evidence is presented in a non-chronological fashion.

In terms of the chronological evidence test, the situation of contemporary scientific research is rather grim. Not only does the scientific community ignore the application of this test, it even insists on applying the exact opposite. Normally one expects that the first theory conceived remains dominant only until it is refuted by another. By contrast, we will see that in the case discussed in this book such a theory may continue to reign supreme even after it has been contradicted in numerous different ways.

This book's scientific advisor has been publishing articles on particle physics, nuclear physics, and electromagnetism since the

1970s. He earned his BSc and MSc from the Hebrew University of Jerusalem in the 1960s and his PhD from Tel Aviv University in the 1970s.

I am a mathematician, a chess problem composer, and a multiple-time world champion in chess problem solving. In my early days I was the winner of the Weizmann Institute's Mathematical Olympics.

I hope you will find this book rewarding and enjoyable.

OFER COMAY

December 2014

The Most Accurate
Theory in History, in Any Field

The Standard Model is a collection of physical theories that describe the most elementary particles in nature as well as the laws that act upon them. The model was developed in the latter half of the twentieth century, and has recently been the subject of great praise thanks to the most expensive experiment in the history of science, which purportedly resulted in the discovery of a Higgs boson, also known as a Higgs particle. The existence of this particle was predicted over fifty years ago and is fundamental to the veracity of the Standard Model.

Despite the majestic aura surrounding this particle, and the fact that one hundred countries had invested billions in experiments that pertain to it, the Standard Model is almost entirely inaccessible to physicists from other scientific disciplines because the Standard Model uses a language and terminology that completely differentiate it from other fields of physics. The language used by the model is highly elaborate, and, because this book rarely employs that language, the criteria by which it can be examined are mainly whether the model accurately predicts experimental data, and what scientists actually say about their own model.

Some Standard Model theories have fared well, such as those that predicted the three generations of particles, but not all Standard Model theories are cut from the same cloth. Some parts of the

Standard Model are indeed self-sufficient and require no reference to other model theories. However, a central part of the model known as quantum chromodynamics (QCD), which describes the structure of protons and neutrons, has consistently failed time and time again.

The theory that describes the structure of protons and neutrons and the force that exists within them, QCD, was developed over forty years ago, and many surprising facts have since come to light concerning what truly transpires inside these particles. Nevertheless, I assume that most particle scientists active today, with the exception of a small group of dissidents,[1] accept the Standard Model as true. For this reason, I should open with an "apology" before I begin my assault on such a universally accepted physical theory. I will now proceed by briefly examining whether the Standard Model of particle physics, with an emphasis on QCD, meets the basic criteria of a well-founded, indisputable physical theory.

Technological Advancements

The discoveries of modern physics have had a tremendous effect on technological progress. The theory of electrodynamics developed in the nineteenth century, and quantum theory, mostly developed in the first half of the twentieth century, now form the basis of almost every device we use on a daily basis. One may con-

---

1   www.physics.auckland.ac.nz/uoa/associate-professor-philip-yock. *"Despite its widespread use, today's Standard Model of physics raises conceptual questions of naturalness, is not free of inconsistencies with observations, and includes unproven conjectures."*

fidently assert that the very existence of these technologies provides a near-absolute validation of the physical theories that underlie them.

It should be noted that currently not a single technology has been developed based on the insights of QCD. I do not consider this to be a weakness of QCD, as it remains entirely unclear whether such knowledge can ever be used for developing new technologies, but it is important to remember that QCD is not supported by any "evidence" in the form of technologies that would have been otherwise impossible to develop.

Five Decimal Places

One of the most appealing arguments in favor of quantum theory and special relativity is the breathtaking precision achieved by their application. Modern particle accelerators are used to collide particle beams with staggering precision. The analysis of collision results is only possible thanks to the perfect correspondence between the mathematical equations of special relativity and the laws of nature as we observe them. In an astonishing feat of empirical evidence, the Large Hadron Collider (LHC) succeeded in providing 10^{15} confirmations of equations taken from special relativity.

The Dirac equation was derived in 1928 and forms a part of quantum theory. The equation predicts real observations with a precision of up to five decimal places. Computational methods developed by quantum field theory have succeeded in achieving even greater precision under certain conditions. This astounding agreement between theoretical predictions and empirical data further supports the veracity of this equation and the computational tools that we employ.

This laudable state of affairs, however, does not apply to QCD. QCD scientists argue that the equations are exceedingly complex, and therefore cannot be applied in low-energy settings.

> We have a simple, definite theory [QCD] that is sup-posed to explain all the properties of protons and neu-trons, yet we can't calculate anything with it, because the mathematics is too hard for us.[2]

This is a legitimate argument, but we should bear in mind that QCD has no "empirical confirmation" that can convince us of its validity.

Different Schools

When dealing with a well-founded physical theory, it is likely to produce the same answers for the same basic questions. The Standard Model, however, offers no conventional answers to even the most fundamental questions. What causes attraction between protons and neutrons? This attraction keeps protons and neutrons together inside the nuclei. That's a pretty basic question, right?

It appears that opinions vary even on such a fundamental ques-tion as this. Distinguished physicists have differing opinions on the matter. Frank Wilczek, a 2004 Nobel Prize laureate, had writ-ten an article in support of the approach that explains the attrac-tion by the interactions of pions and rho mesons[3] that exert the

---

2   Richard Phillips Feynman, *QED: The Strange Theory of Light and Matter*, Princeton University Press, 1985, p. 138.

3   I apologize if readers feel bombarded with new terminology. The issue will be explained later in this book.

force in question and transfer it from protons to neutrons. Other scientists deem this force to be a "residual force" associated with the quarks that comprise protons and neutrons and believe it is unrelated to pions. In one online forum[4] you may find an abundance of contradictory answers given to a naïve student who had been confounded by this basic question.

Salesmen

Strong, well-founded theories do not require the support of writers who lavish praise on them. The Standard Model, however, is backed by a vast legion of authors who regularly commend its success with breathtaking abandon.

When I sat at my computer and googled the phrase:

"Standard Model" most accurate

I was presented with 876,000 hits. Thank God the computer only showed English search results! A significant portion of these websites praise the model's many successes. They also included books, such as *The Standard Model, the unsung triumph of modern physics*[5] (an amusingly self contradicting title). Some websites even depict the Standard Model as "the most accurate and

---

4   www.physicsforums.com/showthread.php?t=382140. *Nuclear force as residual color force.*

5   Robert Oerter, *The Standard Model, the unsung triumph of modern physics*, Pi Press, 2005.

all-encompassing theory in the history of physics,"[6] and the list goes on.

Let's take a closer look at just how accurate these descriptions are. Let's examine Wikipedia's Standard Model article, which describes the accuracy of its predictions (the screenshot below was captured in February 2013):

precision. To give an idea of the success of the SM, the following table compares the measured masses of the W and Z bosons with the masses predicted by the SM:

Quantity	Measured (GeV)	SM prediction (GeV)
Mass of W boson	80.387 ± 0.019	80.390 ± 0.018
Mass of Z boson	91.1876 ± 0.0021	91.1874 ± 0.0021

Figure 1. SM predictions as presented by Wikipedia

The table compares measured (left) and predicted (right) W and Z mass values according to the Standard Model. Even without understanding the role played by these particles, we can still witness a very strong correspondence between predicted and actual

---

6   www.decodedscience.com/the-most-significant-physics-break-through-of-2012/23410.

values. Actual results deviate from predictions by less than one-thousandth! Given that this field of particle physics deals with very high energies, such accurate results are nothing short of fantastic.

There's just one tiny problem: none of that is true. Before 1983, the year when the W and Z particles were discovered, two predictions of their masses were published:[7]

Table 1. Predictions published
before the discovery of the W and Z particles

Particle	Prediction 1	Prediction 2
W	84 ± 2.8	79.5 ± 2.6
Z	94.6 ± 2.3	90 ± 2.1

The predictions were wrong by a few percent. That's quite good, so why push it? By the way, the W and Z particles belong to a section of the Standard Model known as "Electroweak theory." We will refrain from discussing it here, as the validity of this theory neither supports nor refutes QCD theory.

Many proponents tell us that the Standard Model is perfect and devoid of any inconsistencies. Following are a few typical examples (you may find thousands of similar statements online):

---

7   C.H. Llewellyn Smith and J.F. Wheater, Physics Letters **105B**, 275 (1981).

"The Standard Model describes everything we know about the smallest building blocks of nature yet observed. It's the most accurate theory ever developed, in any field."[8]

"[T]he Standard Model has remained fully consistent with all measurements made at the highest-energy particle accelerators to the date of this writing."[9]

"The Standard Model can explain every piece of experimental data concerning subatomic particles up to about 1 trillion electron volts in energy... This is about the limit of the atom smashers currently on line. Consequently, it is no exaggeration to state that the Standard Model is the most successful theory in the history of science."[10]

"Standard Model of Particle Physics is beyond any shadow of doubt one of the biggest accomplishment of Theoretical and Experimental Physics. It is also the most accurate theory ever devised by human beings."[11]

Let's take a quick look at two of the many experiments that refute these bombastic assertions. I've chosen these experiments in particular because they can be visually presented with ease. Notice the five predictions made by QCD, a central part of the Standard Model, that are presented in figure 2. Now look at the actual measurements made in 1983, which are shown in figure 3.[12] According to QCD, the graph should be ascending. In reality, how-

---

8   www.lhc-closer.es/1/6/5/0.

9   Victor J. Stenger, The Comprehensible Cosmos, 2006. p. 92.

10   Michio Kaku, Hyperspace, Oxford University Press, 1995. p. 121.

11   The Standard Model of Particle Physics, by Shahbaz Ahmed Alvi, Department of Physics, University Of Karachi.

12   J.J. Aubert *et al.*, Physics Letters **123B**, 275 (1983).

ever, the graph is descending. This experiment was known as the EMC effect.

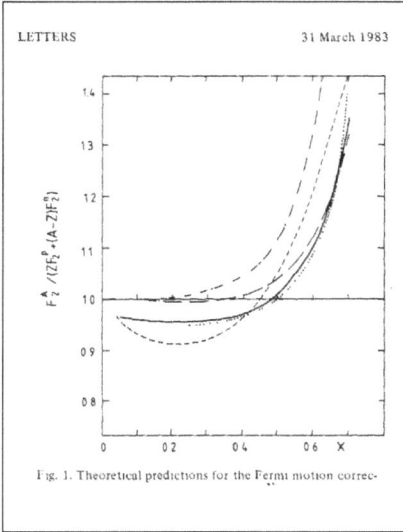

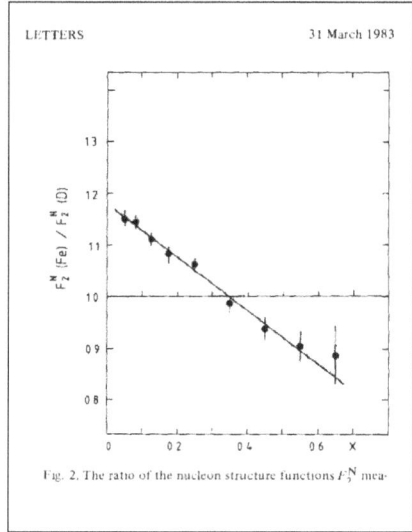

Figure 2. QCD Predictions   Figure 3. Actual Measurements

The results of this experiment, like many others, have not been explained to this day.[13]

The second problematic finding has to do with the alleged discovery of the Higgs boson. The Higgs is a particle whose unique qualities are crucial in explaining the Standard Model. As distinguished CERN physicist John Ellis succinctly put it, "If you see nothing [don't find the Higgs in LHC], in some sense then, we theorists have been talking rubbish for the last 35 years."[14]

---

13   *"So while the experimental signature is clear, the interpretation of this effect is, at present, ambiguous."* Arrington *et al.*, *New Measurements of the EMC Effect in Few-Body Nuclei*, J. Phys. Conference Series **69**, 012024 (2007).

14   *The New York Times*, May 15, 2007.

The CERN LHC experiments did find a new particle. Unstable particles are characterized by the width of their energy. Figure 4 shows what the graph describing the particle's energy width should look like if it were indeed a Higgs boson.[15] In figure 5, however, the actual results are shown.[16] The bell curve generated by the experiment is one thousand (!) times wider than the one predicted by the Higgs boson theory. The difference between projected and actual results is not merely quantitative. The actual results graph shows that the width of the new particle is similar to that of other already known particles that belong to the same mass region, and that it is not special or unique as the Higgs boson ought to be.

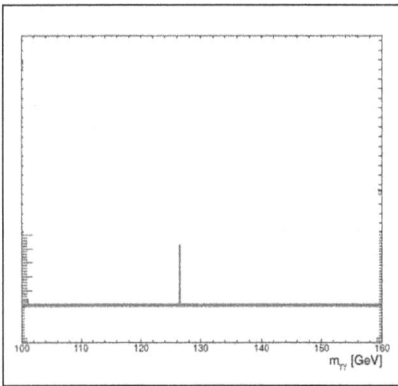

Figure 4. Energy width of the Higgs boson, as predicted by the Standard Model.

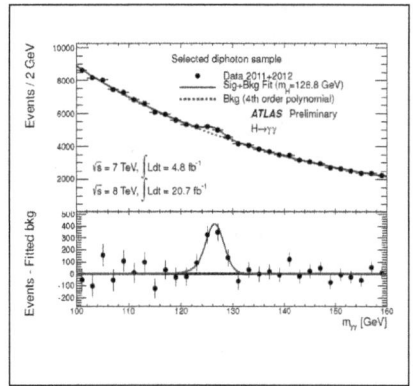

Figure 5. The lower chart depicts actual measurements at the CERN Atlas facility. The curve is 1,000 times wider than predicted.

---

15   Handbook of LHC Higgs cross sections, 17-2-2011, arxiv.org/pdf/1101.0593v3. pdf. p. 143, 145.

16   cds.cern.ch/record/1523698.

It is almost universally agreed that if the particle found in the experiment is not the Higgs boson, then the Higgs boson simply does not exist. There are two options, then: one is that the Higgs boson is indeed a myth and that "we theorists have been talking rubbish for the last 35 years," and the other is that "the most accurate and all-encompassing theory in the history of physics" has made a "tiny" mistake and is off the mark by a factor of one thousand.

By the way, there are scientists who believe that the tremendous discrepancies found in actual data stem from the limitations inherent to particle accelerators,[17] but now is not the time to discuss the issue. We will make do with the fact that these arguments were retroactively invented only after experimental results were finally obtained.

The Agreement between Predicted and Experimental Data

A valid scientific theory should accommodate every experiment that falls within its domain of validity. In principle, a single reproducible experiment that contradicts the theory will suffice in refuting it. Indeed, the stable and valid theories such as special

---

17   *"The total decay width for a light Higgs boson with a mass in the observed range is not expected to be directly observable at the LHC. For the case of the Standard Model the prediction for the total width is about 4 MeV, which is several orders of magnitude smaller than the experimental mass resolution."* pdg.lbl.gov/2013/listings/rpp2013-list-higgs-boson.pdf.

relativity and quantum theory have succeeded time and again in predicting empirical results.

However, there are dozens (!) of experimental data that seemingly contradict QCD. This book addresses around thirty unexplained phenomena, of which more than twenty appear to contradict the Standard Model. These findings have been consistently reproduced time and time again. They are known to the scientific community because they have been published in scientific journals. By the way, there are other data that do not fit the theory, as even QCD proponents have already admitted. I omit them in this book for the sake of brevity and provide references for those who are interested.

How is it possible that in our time, such a basic theory that purports to describe the internal structure of such prevalent particles as protons and neutrons, is in fact completely false? A proton cannot be split open to reveal its inner contents. It is akin to a sealed metal box, and all we can do—even when using the largest particle accelerators to make these boxes collide at vast speeds—is try and predict the pieces that scatter as a result of the collision. The boxes themselves generally return to their original state immediately, and we can examine only the properties of those pieces. If our predictions are correct, we may have a valid theory on our hands. If our predictions are false, we must have made a mistake. Alternatively, it's possible that one of our assumptions is fundamentally wrong.

It therefore follows that it should be rather easy for the scientific community to hold onto an unfounded theory that describes the structure of protons and neutrons. All it needs is the ability to ignore empirical data that contradict the theory, or say that a future experiment may explain them, and go on as though nothing happened. The intention of this book is to allow you to decide for yourself whether particle scientists do, in fact, possess that ability.

The Shock and Awe of Scientists

Testing the validity of QCD requires expensive, protracted experiments that employ gigantic particle accelerators. The majority of these experiments can only be pursued by a few of today's particle accelerators due to the tremendous amount of resources they consume. Consequently, there are long intervals of time separating each experiment, and a decade or even a generation may go by before the next one commences.

Whenever new experimental data contradict QCD, physicists express their candid astonishment. The results are documented but are curiously absent from textbooks. Therefore, when a new generation of physicists encounters new discrepant results, once again genuine astonishment is expressed.

Following is a collection of reactions by scientists to experiments whose results are inconsistent with QCD. The following only address the results that remain unexplained to this day.

> "'[It is a] thorn in the side of QCD,' Sheldon Glashow following an experiment on polarized proton beams conducted in 1979."[18]

> "The results are in complete disagreement with the calculations... We are not aware of any published detailed prediction presently available which can explain the behaviour of these data."[19] (From a scientific article

---

18   *"Glashow once called this experiment 'the thorn in the side of QCD.' In his summary talk at this meeting, Stan Brodsky called this result 'one of the unsolved mysteries of Hadronic Physics.'"* Alan D. Krisch, *Hard collisions of spinning protons: Past, present and future*, The European Physical Journal **A 31**, 417–423 (2007).

19   J.J. Aubert *et al.*, Phys. Lett. **123B**, 275 (1983).

published in 1983 following the discovery of the EMC effect.)

"In 1988, however, physicists were shocked to find experimental evidence suggesting that very little—perhaps none—of the proton's spin comes from the spin of the quarks."[20] (From an article that describes the "proton spin crisis.")

"We have this elegant theory of quantum chromodynamics, which is supposed to describe the binding of the fundamental constituents of all matter, but we don't know how to make it work. We can't even do something as basic as building protons out of quarks."[21] (Robert Jaffe, in a dispirited declaration following a 1989 experiment that failed to detect strange-quark-matter predicted by QCD.)

"This is not an experiment telling us about esoteric things that happened in the first microsecond of the Bing Bang or in some remote part of the universe," says Francis E. Close of the University of Tennessee in Knoxville. "This is the stuff we're made of, and it's showing that maybe we don't understand it as well as we thought."[22]

---

20   Ivars Peterson, Science News, September 6, 1997.
21   Robert L. Jaffe, Science News 1989.
22   Ivars Peterson, *Proton puzzle puts physicists in a whirl*, Science News, April 8, 1989.

"If the results are not a statistical fluke, new physics has been observed. One possibility is that our understanding of what's inside the proton is somehow wrong."[23]

When I wrote to him fifteen years after the experiment, Frank Sciulli wrote back saying that he believes there actually *was* a statistical fluke in his experiment. However, other experiments carried out in the first decade of the twenty-first century using the Tevatron particle accelerator in Illinois have confirmed the existence of an analog finding.

"The results have drastic consequences for the way we understand what the proton is made of," says Charles Perdrisat regarding a 2001 discovery that indicates that the positive charge of protons tends to concentrate in its outer region.[24]

"That's very disturbing. The finding suggests that scientists may have erred in calculations using fundamental theory to predict quark behavior within neutrons." (Xiangdong Ji's remarks on a finding that indicates that quarks have a significant orbital angular momentum.)[25]

---

23   Frank Sciulli, Columbia University News, 1997.

24   Peter Weiss, *New probe reveals unfamiliar inner proton,* Science News, May 5, 2001. Another citation from this article: *Theorist Ulf G. Meissner of the National Research Center in Jülich, Germany, admits to being stumped by the new results and wonders if they will hold up. He says he knows of no model of the proton that would lead to such bizarre distributions of the electric field. Says Meissner, "For me, it's a real puzzle."*

25   Peter Weiss, *Topsy Turvy: In neutrons and protons, quarks take wrong turns,* Science News, vol 165, 2004.

It's interesting to see how Frank Wilczek, one of the pioneers of QCD, addresses the incongruence of QCD and known phenomena:

> "Ironically, from the perspective of QCD, the foundations of nuclear physics appear distinctly unsound."[26]

In this article, Frank Wilczek explains that this theory is incompatible with some of the fundamental properties of atomic nuclei, and that he hopes a breakthrough will be achieved later that will resolve the issue.

Surprisingly, Wilczek fails to convey any astonishment at QCD's disagreement with reality.

---

26   Frank Wilczek, *Hard-core revelations*, Nature, **445** 156 (2007).

The Cave Allegory

In my correspondence and meetings with scientists active in the field, I have been surprised to realize that the full extent and the context of QCD discrepancies are unknown to many of them. The allegory I present below perfectly corresponds to a large portion of the topics discussed in this book, and to the internal logic that underlies each topic and that bridges the gap between them.

Discovery and Crisis

In the 1960s a giant cave was discovered in the Sahara Desert. The walls of this cave were covered with magnificent paintings drawn some eight thousand years ago. The paintings depicted individuals who were exceedingly tall.

The discovery had caused a palpable stir within the scientific community. It was clear that a new civilization had been found. But things didn't add up. The scientists made two assumptions:

1. There was no available water in the region. The cave was located deep in the world's largest desert.

2. The paintings depicted exceedingly tall individuals. It was estimated by scientists that some of them were approximately seven feet tall. Humans of this size could not have survived without plenty of food and water.

The new finding was troubling to scientists at the time, and it was known in textbooks as the sixties crisis. But in 1972, a new revolutionary idea began percolating into scientific literature. A group of scientists had demonstrated that there is a specific DNA sequence that produces exceedingly tall, highly intelligent living creatures who require no water at all to survive. The opposite is actually true—creatures with such DNA could not have survived next to a water source. The theory was known as Quasi Chain DNA (QCD).

Sexual Freedom

Despite the fact that the theory was completely unprecedented in the realm of natural science, one of the scientists who proposed this theory was a leading figure in the field. It is for this reason that scientists decided to give the theory a chance and see whether other findings extracted from the cave were compatible with the theory.

The wall paintings showed a great distance between the figures they contained. When scientists examined the genetic sequences of QCD, they discovered that creatures that have such sequences prefer to engage in intercourse from a distance, the opposite of any other mammal known to science.

This special property of QCD creatures is known in literature as "sexual freedom," and is considered to be evidence that supports the validity of QCD to this day. In 2004 a Nobel Prize was awarded to the three scientists who revealed sexual freedom.

It wasn't long before the new theory garnered additional achievement. QCD scientists predicted that these creatures would prefer to live in three-member families. In 1978, an additional cave was found in the same region that contained new paintings. In those paintings there are three figures dancing, probably two parents with their child. These splendid paintings, known as the Three

Dancer Event, are described as powerful evidence for QCD, even though human creatures with ordinary DNA behave in a similar fashion.

Contradictions

Even the earliest excavations revealed tools that were staggeringly identical to other agricultural tools found in North Africa. The resemblance remains unaccounted for to this day. Leading scientists in this field have said that these tools appear to contradict QCD.

In 1974 several pools were found near the cave. This discovery shocked scientists, since QCD creatures were unable to survive near water. However, scientists defended their theory by saying that the pools never contained any water.[27]

According to QCD, the unique genetic sequence of QCD individuals produces equally tall males and females. However, paintings found in the cave and in other caves in the region show that men were statistically taller than women. This finding was published in 1977, and its discoverers were sent to collect statistical data from other caves in order to corroborate it. In the 1980s additional findings were uncovered in other caves, indicating that, similar to humans, QCD males were also taller than females. However, QCD proponents argued that this discrepancy doesn't preclude the possibility of future evidence that supports the theory.

The unique genetic sequences of QCD were also able to predict the social behavior of these creatures. Based on those sequences, scientists predicted that these creatures did not require gods or

---

27   Later in this book readers will find a real-world analogy that shows that the comparison is not excessive, and that no attempts are made to disparage scientists.

kings, as opposed to other human creatures. However, an independent group of researchers known as the Explorers of Mountain Caves (EMC) had discovered a palace and a shrine-like structure near the cave in 1983. This discovery flew in the face of predictions published before it came to light, and it remains unexplained to this day.

In 1987 EMC scientists had found additional caves with bedrooms and beds that were used by the QCD cave people. The size of the beds indicated that the creatures weren't tall at all, and were in fact only slightly taller than five feet, a height similar to that of other humans living during that period. This discovery also remains unexplained to this day and appears to contradict QCD predictions.

In the 1990s the remnants of microorganisms that had lived inside the pools were found. This discovery indicated that the pools did in fact contain water. In the beginning of the twenty-first century, the pools also revealed the remains of fish skeletons. This provided irrefutable evidence that the pools did in fact contain water. The discovery remains unexplained to this day since the theory states that QCD creatures were unable to survive near water.

As years went by, many additional caves were discovered, allowing scientists to conduct statistical analyses. It was found that larger families had lived in smaller dwellings. This finding appeared to contradict the sexual freedom principle, as it indicated that many sexual interactions took place between creatures living in rather crowded spaces, as is the case for ordinary human beings, and therefore it contradicts QCD theory. Another troubling finding was revealed when paintings that showed two figures, most likely from different families, engaged in sexual intercourse. The paintings show that this interaction intensified the closer the figures were to each other. It also appears to contradict the behavior expected of QCD creatures, who supposedly enjoyed sexual freedom—namely, their sexual interaction should have diminished as they became closer to one another.

QCD experts meticulously analyzed the DNA sequences of QCD creatures and concluded that these creatures had skeletons that differed from those of humans. Based on this finding, scientists predicted that certain non-human-like bones would be found, known in literature as "exotic bones." However, to this day no exotic bones have been discovered, despite extensive efforts to uncover them.

Several years before the cave discovery, scientists found channels in the region that looked like part of an irrigation system. Scientists adopted a theory called Violent Meteor Dominance (VMD), which argued that the channels had been formed by a meteor shower. VMD was taken seriously for a few decades, despite the fact that a meteor shower cannot produce straight channels. Later the VMD explanation was abandoned and removed from textbooks. The channel phenomenon was removed as well, probably because it cannot be explained by QCD.

In Summary

So what did we have here? Let's begin with the two assumptions that led to the aforementioned discrepancies made in the 1960s:

1. There was no water near the caves.

2. The creatures were very tall.

Both of these assumptions are now highly suspect. These were the assumptions that led scientists to believe that the creatures weren't human, thus inspiring the concept of QCD.

Moreover, as years went by every finding revealed in other caves exposed in the same region has shown that these creatures were similar to ordinary human beings. This appears to contradict QCD theory, which assumes that the characteristics of these creatures were much different from those of human beings.

What Scientists Say

Scientists continue to regard QCD as the only possible theory. When confronted with contradictory evidence, they say that the mathematics of these creatures' DNA sequences is so elaborate that it is likely that even the fastest computers in the world would be unable to ever accurately calculate their properties.

Furthermore, they argue, QCD has been very successful in predicting the growth rate of this desert population. For this reason, they believe, QCD must be true. By the way, there are other models that do not assume the validity of QCD that also achieve accurate growth rate calculations.

Today, mainstream scientists are very happy with QCD theory. Some argue that even if the future fails to reveal explanations that resolve the issues presented above, they would nevertheless avoid examining any new theories. We simply have to live with the knowledge that, since QCD sequences are so complex, we'll never be able to calculate the properties derived from them with satisfactory precision.

And that, in a nutshell, is QCD theory.

A series of events that are almost perfectly analogous to the cave allegory took place in the exact same period in the field of particle physics. In the epilog of this book you will find the cave allegory retold in almost the exact same words, only this time the allegory applies to physics. Those of you who succeed in following the narrative of this book will note that the particle physics version of this allegory is even more curious.

THE BIG SIXTIES CRISIS

Why Everything Is Arranged in Shells

I suppose that at this point the readers of this book who happen to be particle physicists are angry enough to demand that I begin substantiating my claims. However, we will have to learn quite a bit so that we can not only hear the claims, but also understand them and even take our own stand. The purpose of this book is to provide readers with tools that will enable them to truly comprehend the subject, as opposed to merely exposing them to arguments that relate to a certain consensual theory, without being able to judge for themselves whether or not the arguments actually make sense.

QCD was born out of a crisis that began in the 1960s, with the discovery of the particles that comprise protons and neutrons. Physicists at the time thought that it was impossible to use known physical laws in order to describe them, and so they had to invent a whole new form of physics.

The founders of QCD introduced several unprecedented assumptions, such as forces that operate through three different entities known as colors, particles that can't exist unless they contain an equal portion of all three colors, equations indicating that the forces intensify as particles become more distant from one another, and other such fantastic properties hitherto unknown to nature. Such a fantastic theory can only be reasonable if the tried-and-true physics of the twentieth century completely fails in describing the inner workings of the proton.

Therefore, we will begin by learning more about that "tried-and-true physics" on which almost all of the technologies we use today are based.

In this chapter and the next we will briefly examine all of the topics that are relevant to understanding the crisis. Most of the material in these two chapters will be familiar to anyone with a bachelor's degree in physics.

These chapters are rather technical, and readers who feel they need not consider the matter too deeply can choose to quickly skim or flip through these pages and take a quick look at the illustrations I have endeavored to supply in abundance. For readers who have never encountered such materials, it is recommended that they regard the data appearing in the following pages as "ground rules" to be applied later.

The Theory of Electrodynamics, Wave Theory, and Special Relativity

In 1861, Maxwell formulated the theory of electrodynamics that take the form of several equations known as Maxwell's equations. Maxwell used known formulas and introduced his own amendments to them. According to the theory of electricity there are positive and negative charges. Charges of equal sign repel, charges of opposite signs attract, and light beams interact with both charges. When a beam of light is emitted from the sun and shines on a cat enjoying its warmth, the light transfers its energy to the electrons located in the atoms of our cat, and is thereby absorbed by it.

In 1900 Planck published his own equation according to which light is absorbed discretely rather than continuously. The Planck equation describes the relationship between the absorbed energy of light and its wavelength. The proportion between the two is a constant number known as the Planck constant. Planck did not ascribe a great deal of significance to his equation,[28] which is considered to mark the birth of quantum theory.

In 1905 Einstein published several essays that dramatically transformed the field of physics as it was known at the time. His first essay[29] concerned the photoelectric effect, and in it Einstein for the first time proposed the idea that light is not a continuous flux of energy, but rather it is composed of discrete quanta. Each such quantum is a particle known today as a photon. According to Einstein, electrons can only be arranged in certain ways whenever they are held inside an atom, and, if a single quantum of energy is unable to change the position of the electron, it follows then that the electron and photon fail to interact altogether.

Today Einstein's and Planck's ideas seem completely natural, but at the time they were encountered by fierce resistance, and it took another decade before they were accepted by the scientific community.

Figure 6. When a photon (γ)[30] lacks sufficient energy to shift the atom to a higher energy level, it passes unperturbed.

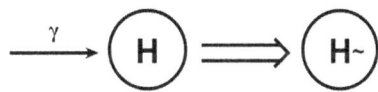

Figure 7. When a photon possesses enough energy, it is absorbed by the atom and shifts it to a higher energy level.

---

28 Helge Kragh, *Max Planck: the reluctant revolutionary*, Physics World. December 2000.

29 A Einstein, *Ist die Trägheit eines Körpers von seinem Energieinhalt abhängig?*, Annalen der Physik, 1905.

30 The "γ" symbol in this book denotes a photon with an appropriate energy, and not necessarily very energetic γ ray.

Another publication by Einstein that same year provided the foundation of special relativity theory.[31] Special relativity is tremendously important for understanding the universe in general. For our purposes, we briefly summarize the relevant topics as follows:

- The speed of light is constant. This result can also be obtained by using Maxwell's equations. This is very surprising, as it contradicts both Newton's laws and our own intuition.

- Every particle in the universe belongs in only one of two groups: those that always travel at the speed of light, and those that always travel at a slower-than-light speed.

- The mass of an object is proportional to the total amount of energy it contains. Therefore, if, for example, we allow an object to accelerate, we also provide it with speed and energy, thus increasing its mass.

According to special relativity equations, the mass of an accelerating object grows dramatically whenever it approaches light speed. If its speed is significantly lower than the speed of light, then its change in mass is almost imperceptible. This mass-changing effect is known as a "relativistic effect," and it is habitually disregarded when we use approximate equations that concern particles that travel at much slower-than-light speeds.

Spin and the Pauli Exclusion Principle

By exploiting the fact that the wavelength or the frequency of light enables us to measure the energy of a single photon, and the knowledge that electrons emit or absorb photons and change their

---

31   A Einstein, *Zur Elektrodynamik bewegter Körper*, Annalen der Physik, 1905.

energy levels inside the atom, scientists can then measure the energy levels of electrons held inside atoms. Naturally, research efforts were at first focused on the simplest atom of all: the hydrogen atom, which contains a single proton and a single electron.

Scientists were looking for a theoretical explanation for the findings they uncovered on the energy levels of the electron inside a hydrogen atom.

The favorite model of the 1910s was that of Niels Bohr, although this model is no longer accepted. The model is sometimes referred to as the "first quantum theory." One of the most significant physical quantities already used by the first quantum theory was the h-bar ($\hbar$), which equals the Planck constant divided by 2π. The $\hbar$ constant is a unit of angular momentum obtained by rotating objects. Even during the days of the first quantum theory it was possible to measure the intrinsic angular momentum of the electron, and it was revealed that it is equal to $\frac{1}{2}\hbar$. Physicists describe this property in an abbreviated fashion by noting that the electron "has spin $\frac{1}{2}$." Particles with half-integer spin are known as fermions, named after the Italian physicist Enrico Fermi. The proton's spin was also measured, and it was discovered that the proton is also a fermion, and its spin is equal to $\frac{1}{2}$.

Let's now fast forward to the 1920s, a period during which quantum theory as we know it today had already flourished.

In 1925 Wolfgang Pauli formulated a principle, the Pauli exclusion principle, which, in simple terms, states that two fermions of the same type can never be in the same quantum state at the same time. The intuitive implication of this is that a fermion is a particle that occupies space, and therefore it is impossible to have two identical fermions that are simultaneously located in the same place. The Pauli principle is one of the most fundamental and significant principles in nature. When we clap our hands, for example, if it weren't for the Pauli principle, our hands would simply go through each other uninterrupted. It is most fortunate, then, that Pauli went through all that trouble and came up with his principle!

It is interesting to note that it wasn't until the 1960s that the Pauli principle was proven to be what prevents objects from passing through each other.[32]

The Schrödinger Equation

In 1926 Schrödinger developed a new equation, the solutions of which aptly described the possible states of the electron situated inside a hydrogen atom. An electron is both a wave and a particle, and the Schrödinger equation enabled us to describe the electron's wave function.

An examination of the equation's solutions revealed that electrons can be situated at the lowest energy level, known as the "ground state." When the electron is at the lowest energy level, the hydrogen atom's wave function displays a spherical symmetry—namely, the electron can be located, with identical probability, at every location surrounding the proton, and this probability depends only on the distance of any given point from the proton. This solution is known as 1s. Another solution relates to an electron that also exhibits spherical symmetry, and that is situated farther from the center, and is thus located at a higher energy level. This solution is known as 2s. Spherical symmetry solutions are known as s-wave solutions.

---

32   F.J. Dyson, A. Lenard, *Stability of matter*, Journal of Mathematical Physics, 1967.

The Latin letters that describe the waves were chosen for historical reasons about which I won't trouble readers.

The equation has other solutions that describe additional electron states. These solutions are not characterized by spherical symmetry. The next possible solution is called a p-wave. A p-wave solution at the lowest energy level is known as 2p. There are, in fact, three 2p solutions at the same energy level, and each of these solutions is characterized by the same asymmetry, only in a different direction. The next p-wave solution is known as 3p, and so on. The next solution type is called a d-wave, and so on. The Schrödinger equation tells us that the energy level of a 2s orbital is equal to that of 2p, the energy level of a 3s orbital is equal to that of 3p, and so on.

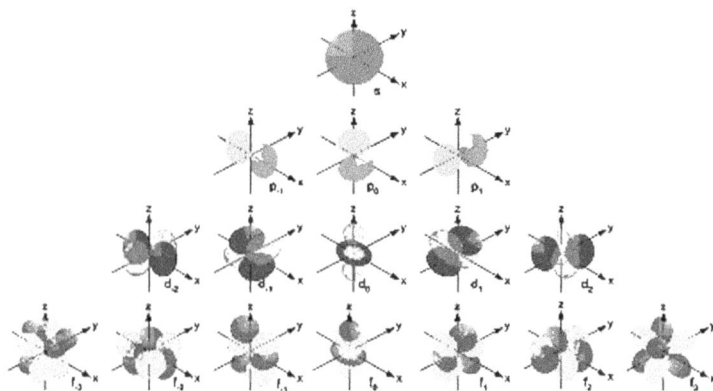

Figure 8. Excerpted from a basic chemistry textbook, this figure describes all possible electron states inside a hydrogen atom. There is a single s-wave orbital (first row), three p-wave orbitals (second row), and so on.[33]

---

33 chemwiki.ucdavis.edu/Wikitexts/UCD_Chem_2A/ChemWiki_Module_Topics/ Electrons_in_Atoms/Electronic_Orbitals.

What is the meaning of all this? As far as the hydrogen atom is concerned, all it tells us is that these are the possible states and energy levels of its electron. Each of these states actually describes two possible electron states, one characterized by a $+\frac{1}{2}$ spin and another with a $-\frac{1}{2}$ spin. We will soon see how this distinction manifests itself.

Why Are Electrons Arranged in Shells?

The Schrödinger equation also applies to atoms that contain multiple protons in their nuclei. When you apply the Schrödinger equation to atoms that have a nucleus with a greater positive charge, and therefore a larger number of electrons, you discover that each electron is situated in a kind of shell, and that each such shell is characterized by a certain energy level. This arrangement in shells stems from the fact that all electrons are fermions that comply with the Pauli principle and that are situated in a certain place—namely, it is impossible to have two electrons in the same state at the same time.

Simply (and inaccurately) put, as we will see later, in a helium atom, which contains two protons and two electrons, and which is situated at its lowest energy level, we have two 1s electrons with two opposite spins. These opposite spins do not interfere with one another as far as the Pauli principle is concerned, as they do not occupy the same quantum state since they have two opposite spins.

Every electron state at a certain energy level is referred to as a shell. Some distinguish between the 2s and 2p shells, but for the sake of this discussion we will regard a 2s electron as though it is located in the same shell as a 2p electron, since the two electrons have approximately the same energy level. We say that a shell is full when all the positions in a certain shell are occupied.

According to this arrangement, when a shell becomes full, and the next electron needs to go to a higher-energy shell, the atom possesses relatively low energy. An atom with full shells is usually referred to as a "noble gas." This description provides a good explanation as to why certain atoms function as noble gases—namely, they are stable atoms that do not need to bond with other atoms in order to reduce their energy level.

In a lithium atom, which has three electrons, the third electron cannot be in the same shell as the other two due to the Pauli principle, and so it occupies the 2s shell. Therefore, the lithium atom has a higher energy level than the helium atom, and it tends to form compounds in order to reduce its energy level. The next noble gas is neon, which has ten protons and ten electrons: two 1s electrons, two 2s electrons and six 2p electrons. None of these electrons interferes with the others owing to the Pauli principle, as each occupies a different quantum state.

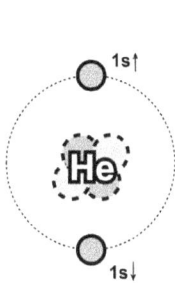

Figure 9. Two helium atom electrons occupying an 1s shell with opposite spins.

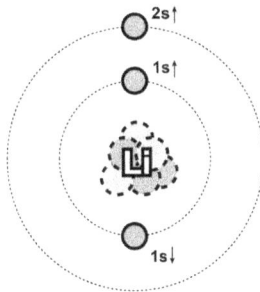

Figure 10. The third electron in a lithium atom cannot be located in the inner shell, and therefore occupies a higher shell.

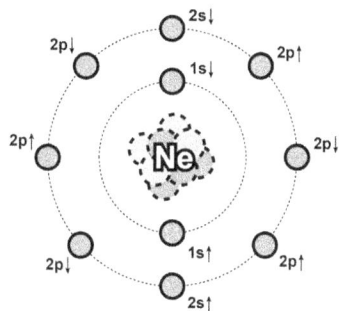

Figure 11. Ten electrons inside a neon atom, a noble gas. The second shell contains two 2s electrons and six 2p electrons.

Evidently, this is how we explain the electrons' arrangement in shells. The higher the energy of the electron, the higher the shell in which it is located.

As I mentioned earlier, this description is not entirely accurate, and I provide a more rigorous explanation later.

Other Types of Shells

The principle behind the Schrödinger equation can be used not only for describing the structure of atoms, but also for describing other states of particles made of multiple bonded fermions.

Let's examine the structure of an atomic nucleus. The nucleus contains both protons and neutrons, collectively referred to as nucleons. Schematic and simplistic figures occasionally show them to be attached to each other as though they were superglued together, but in fact they are moving faster than the electrons of the same atom.

The atomic nucleus does not have a center that attracts nucleons. The nucleons attract each other, and the force that attracts them to each other is called the "nuclear force."

Protons and neutrons, being fermions themselves, must also comply with the Pauli principle, and they too are arranged in shells in a similar fashion to those of the electrons inside the atom. An ordinary hydrogen nucleus has only one proton, but the common helium nucleus has two protons and two neutrons. Both protons and both neutrons occupy the 1s state, and the spin of each proton/neutron is opposite to the other's. This enables all four fermions to comply with the Pauli principle, despite the fact that they all occupy the 1s shell. The protons do not interfere with each other because they have opposite spins, and the protons do not interfere with the neutrons because they are not the same type of fermion. Indeed, the helium nucleus, composed of two protons and two electrons, has a particularly low level of energy.

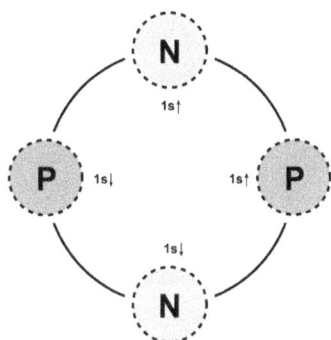

Figure 12. Two protons and two neutrons inside the inner shell of a helium nucleus (also known as an α particle). The neutrons do not interfere with one another because they have opposite spins, and the protons do not interfere with each other because they too have opposite spins. Finally, protons do not interfere with neutrons since they constitute different types of fermions.

Larger nuclei comply with similar principles. Just as there are noble gas atoms with lower energy levels and thus greater relative stability, there are also nuclei with closed, relatively stable shells. The number of protons and neutrons in such closed shells are called "magic numbers."

When it was discovered that the proton (and neutron) are actually composed of quarks, it turned out that quarks also have half-integer spins, and are thus fermions themselves. That immediately raised the question of how we could apply the Pauli principle and the shell arrangement phenomenon to describe the inner structure of the proton. This key question was first raised at some point during the 1960s and resulted in the formulation of QCD. We will soon consider this question at length, but let's first continue examining the "ground rules" for just a while longer.

Spin and Parity

This chapter is devoted entirely to the concept of spin and parity, and, in particular, it explains how an atom's spin is calculated. The topics discussed are not simple, but careful readers will acquire tools that will later enable them to:

- Understand the conflict that resulted in the formulation of QCD.

- Understand the proton spin crisis, one of the most important mysteries of physics.

- Refute a common error found in numerous basic textbooks, an error many physicists are unaware of.

Spin and Orbital Angular Momentum

Closed systems are characterized by conservation laws, such as the law of conservation of energy. There are other quantities governed by such conservation laws, one of which is angular momentum.

Even a particle at rest may have angular momentum. The angular momentum of a particle at rest is called spin, and you can visualize it as the motion of a spinning top that spins in place. The spin of a particle that consists of multiple smaller particles is equal to the sum of the angular momentum of all its constituent particles.

If we want, for example, to calculate the spin of a helium atom, we will have to combine the angular momentum of its two electrons. For example, let's assume that they both occupy a 1s shell, and that they both have opposite spins. Under these circumstances, the two spins will cancel each other out, and the helium atom's total spin will be zero. This form of addition is called "vector addition."

An electron occupying another shell, say, 2p, has another rotational component in addition to its spin. The motion[34] of such an electron cannot be depicted by a symmetrical sphere, and this asymmetry results in an angular momentum that is ascribed to the entire atom. The contribution of such a p-wave would be an integer equal to either +1, 0, or -1. Moreover, we must also add the spin of the electron itself, which is equal to either +½ or -½. Each such number is actually a multiple of $\hbar$, which is usually omitted for the sake of brevity.

Consequently, all solutions apart from the s-wave result in orbital angular momentum, sometimes referred to as spatial angular momentum. In order to calculate the total spin of the atom, we must add together everything that contributes to angular momentum: the spin of all electrons and their orbital momentum. Each quantity can be either negative or positive.

An electron found in a higher solution of the Schrödinger equation, e.g., a 2d shell, still has an integer orbital momentum, although in this case the spin ranges between -2 and +2. The electron in the next shell, known as 3f, has an orbital momentum that ranges between -3 and +3, and so on.

Notice that angular momentum, as is the case for electric charge, is quantized; namely, there is a minimal angular momentum equal to $\hbar/2$, and the angular momentum of any particle, either elementary or composite, is a multiple of this quantity by an integer. Or-

---

34   I hope physicists can forgive me for using the term "motion" and "sphere" rather than "wave function."

bital angular momentum, however, will always be an integer multiple of ħ. Quantum physics correctly describes this experimental result.

Parity

Parity is an interesting quantity that we should be able to use, since QCD fails to employ it properly as we will later witness ourselves. The description brought to you here is technical in nature, but it is simple enough for us to use it later on.

An electron occupying an s shell is even, whereas a p shell is odd, a d shell is even, an f shell is odd and so on. The parity of the entire atom depends on the parity of the single electrons that comprise it. We say an atom is in an even state if it has an even number of electrons that occupy an odd state. Feeling confused? It's alright. I also had to read the sentence I just wrote five times before I could understand it myself. For example, the ground state of the carbon atom has six electrons: two 1s electrons, two 2s electrons, and two 2p electrons. Hence, due to the even number of 2p electrons, the ground state of the carbon atom is even.

Parity is an important quantity, as it turns out that an atom maintains its parity when not affected by external factors. The photon, a particle that constitutes a single quantum of light, has a spin of 1 and an odd parity. It turns out that when a photon collides with an electron situated in an atomic shell and is absorbed by it, the atom's parity is consequently altered. If the atom was originally even, it now becomes odd, and vice versa.

I'm assuming that those readers who are first encountering the concept of parity in this book will likely not really understand what it means. As we will see later, parity and parity conservation are basic physical principles, the comprehension of which will aid us in understanding the many flaws found in QCD. At this moment, however, we will consider the whole thing as nothing more than a fun game, and one of the rules of that game is that parity is always conserved.[35]

The Spin of Closed Shells

When shells are full (or closed), as is the case of noble gas atoms that are at their lowest energy level, the electrons occupy every possible state in the atom's closed shells. In this case the sum of electron spins is zero, as they cancel each other out, and the angular momentum is also equal to zero because it too is cancelled out.

Non-noble atoms, however, have a closed shell in their center, where the total spin is equal to zero, and thus it can be said, with good approximation, that only the spin and angular momentum of those electrons occupying the outer shell provide any contribution to the entire atom's spin.

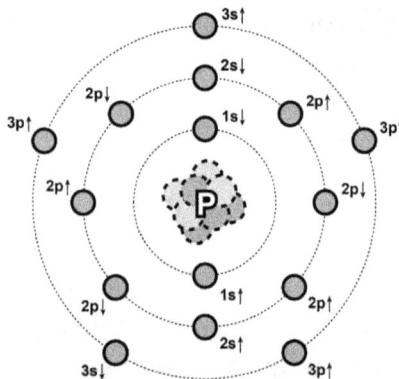

Figure 13. The ground state of a phosphorus atom with ten electrons in its inner shells, with a total angular momentum of zero. Only the five electrons occupying the outer shell determine the phosphorus atom's total atomic spin.

---

35   Unless weak interactions are involved. Weak interactions are not the main subject of this book.

Hund's Rules

The Schrödinger equation elegantly describes the forces between a hydrogen atom and its electron, and until now we have only discussed larger atoms in terms of the attractive force that operates between electrons and the nucleus, while disregarding the repulsive force between electrons.

As we will see by the end of this chapter, the repulsive force between electrons is crucial for our ability to understand the structure of atomic shells. I am convinced that had particle scientists addressed such forces as well, and applied the insights that stem from them to the structure of nucleons, QCD would never have seen the light of day.

We will start with the relatively simple concept of Hund's rules.

In the 1920s a scientist by the name of Friedrich Hund examined the behavior of multi-shell atoms, particularly the sort of states that are "preferred" by electrons in the outer shell when the atom is in its ground state. He discovered several rules that are now named after him, Hund's rules. The most famous of these rules concerns the direction of electron spins. This rule dictates that as long as the Pauli principle is maintained, electrons tend to arrange themselves so that all of the spins in the atom are parallel to one another.

Why is that so? The conventional explanation of Hund's rules says that when electron spins are facing the same direction, their effective distance is greater, which is preferable because the repulsive force between the electrons is thus minimized. The greater the effective distance between the electrons, the weaker the repulsive force between them, and the lower their energy levels.

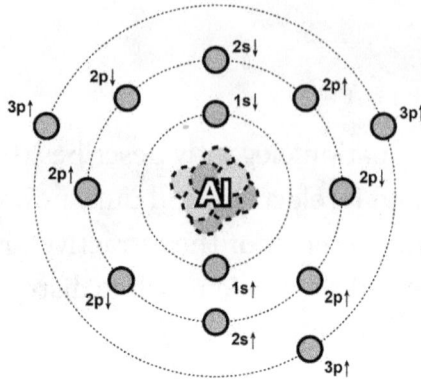

Figure 14. Hund's rules applied to the ground state of an aluminum atom with thirteen electrons. The three outer-shell electrons (3p) all have spins in the same direction.

The Algebra of Wigner and Racah

According to what we have learned so far, the arrangement of electrons in shells is rather technical, but not overly complicated. This is how the subject is taught in universities and that is how most physicists understand it.

This simplistic representation, however, is very far from reality, and is only accurate when it comes to the hydrogen atom. It appears that the forces that operate between electrons provide a crucial contribution to our understanding of the true state of affairs. In fact, the reality is so complex, that almost all physicists choose to ignore it, and instead use the relatively simplistic concepts that have been described up to this point.

Immediately after the publication of the Schrödinger equation, the energy levels of larger atoms were measured, which turned out to be incommensurate with the equation. Physicists have tried to develop mathematical models that could aptly describe the situation, and they have invented methods such as the Hartee-Fock Method or the Thomas-Fermi Method, but these equations offer only approximations, and they assume that the electrons are indeed arranged according to the shell principle described earlier, with the addition of several small corrections based on other forces that operate between electrons inside the atom.

An accurate description of atomic shells will require us to acquaint ourselves with a new concept: configurations.

Our atomic descriptions have thus far involved a possible arrangement of electrons inside the atomic shell. One such arrangement is called a configuration. As a matter of fact, each atomic state is described as a superposition of configurations. In simple terms, one can say that the forces that operate between the electrons cause them to constantly shift from one configuration to the next, while maintaining the atom's total spin and parity. In other words, each atom actually contains a collection of configurations, each having the same spin and parity. The hydrogen atom is the only one that has a single configuration, as it contains only one electron. Configurations appear in atomic shells whenever repulsive forces exist that operate between electrons.

Physicists had already recognized the significance of configurations sixty years ago, not only when it comes to large atoms, but also when dealing with small, non-hydrogen atoms such as helium.[36]

Complicated, right? That's what physicists think, too, even though this theory had generated spectacularly accurate results when describing the characteristics of many atoms, and these

---

36   G.R. Taylor and R.G. Parr, *Superposition of configurations: The helium atom*, Proc., Natl. Acad. Sci. USA **38**, (1952). p. 154–160.

are sometimes accurate up to six decimal places! The algebra utilized in a calculation of atomic configurations is very complex, despite the fact that in the 1940s Eugene Wigner (a Nobel Prize laureate) and Yoel (Giulio) Racah (Israel Prize laureate) separately developed an ingenious mathematical method that makes these elaborate calculations easier to complete. This complexity has led many physicists to engage in many painful compromises with reality and to declare that total accuracy is unnecessary. An approximation of the truth will suffice; namely, the simplistic theory mentioned previously that disregards multiple configuration states.

Indeed, even textbooks[37],[38] and websites[39],[40],[41] that contain educational materials describing the arrangement of electrons inside a helium atom, for example, include only the simplistic portrayal of the helium atom as containing two electrons with opposite, 1s spins, without ever mentioning that there are other possible configurations.

The calculations made with the appearance of the first computers (in the late 1950s) have shown that atoms with multiple electrons have many possible configurations, and in fact are rarely found in their basic configuration![42] Therefore, the approximations used by physicists who ignore this theory are very far from reality.

---

37   John Olmsted, Gregory M. Williams, *Chemistry, the molecular science*, (John Wiley & Sons, Inc), p. 310.

38   Frank H. Shu, *The physical universe: an introduction to astronomy*, p. 51.

39   hyperphysics.phy-astr.gsu.edu/hbase/quantum/helium.html.

40   wiki.brown.edu/confluence/download/attachments/29133/Helium+ and+Calcium.pdf.

41   quantummechanics.ucsd.edu/ph130a/130_notes/node35.html.

42   A. W. Weiss, *Configuration Interaction in Simple Atomic Systems*, Phys. Rev. **122**, (1961). p. 1826–1836.

A solid conclusion of the configuration concept is that bonded particles that contains three fermions or more will never occupy a pure s-wave state, and that a significant portion of angular momentum is orbital. Later we will see that the erroneous assumptions of physicists whereby the proton and other particles are made of quarks that always occupy an s-wave state are what eventually caused the crisis of the sixties and ultimately led to the formulation of QCD. These assumptions are also the reason for the proton spin crisis, a mystery that remains unsolved to this day.

Yoel Racah further developed this theory and demonstrated that it can also be used to describe the shells inside atomic nuclei. When quarks were discovered in the 1960s, one would assume that Racah continued developing his theory so that it would apply to them as well. Unfortunately, in 1965 Racah died in an accident involving a broken bathroom gas tap, and Wigner discontinued his work in this field around the same time. And so, despite the tremendous significance of Wigner and Racah's theory to understanding the structure of atomic shells, there were no notable scientists at the time who pursued its further development.

I conclude this chapter with a little story to illustrate how this important theory was cast into oblivion. The Jerusalem-based physicist Nissan Zeldes, one of Racah's disciples, said that, to mark Racah's one-hundredth birthday, he sent an article that summarized his work to a leading journal. The editor told him that he had no idea what the article was about, and that he was unfamiliar with any scientist working for his journal who could say whether or not the article was worthy of publication. At the end of the day, the article was published elsewhere.[43]

---

43   Nissan Zeldes, *Giulio Racah and theoretical physics in Jerusalem*, Archive for History of Exact Sciences , Volume **63** (3) Springer Journals, May 1, 2009.

From Dirac to the Sixties Crisis

In this chapter we will examine the scientific development achieved since the discovery of the Schrödinger equation and until the discovery of quarks, as well as the crisis that occurred as a result during the 1960s.

Those who have carefully read the previous two chapters will be able to understand the conflict that gave rise to the QCD theory, and maybe even resolve it on their own.

The Dirac Equation

Despite its being practical and intuitive, the Schrödinger equation has two theoretical limitations. First, the equation is not relativistic; namely, it does not take into account the change in mass that is brought about by the electron's speed, and so it only serves as an approximation of a more accurate equation. Schrödinger's approximation is fairly good, since the electron's speed inside the hydrogen atom is considerably slower than the speed of light. Another limitation stems from the fact that it does not address the electron's spin. The concept of spin was artificially combined with the equation's solutions in order to describe all possible electron states inside the hydrogen atom.

In 1927, a year after the Schrödinger equation was published, Pauli demonstrated how it's possible to formulate equations that also address the electron's spin. A year later Dirac formulated his own Dirac equation, the structure of which was consistent with special relativity. It turns out that the equation describes an elementary particle (namely, a particle not composed of smaller constituent particles) with a spin of ½. According to Dirac, this spin ½ property is the result of a mathematical analysis of the equation, rather than an assumption on which the equation is based. In other words, the equation explains the spin ½ property of electrons and of other particles.

The Dirac equation is considered one of the pinnacles of quantum theory. It surprisingly demonstrated that for each particle for which the equation holds (now known as a Dirac particle) there is an "antiparticle" with identical properties, but with an opposite charge. And so, a short while later, the existence of a positron—an electron with a positive charge—was confirmed. To this day, almost all elementary particles discovered have been Dirac particles.

Antimatter

When a particle and an antiparticle meet (such as an electron and positron), they destroy each other and produce an amount of energy equal to that contained in the mass of both original particles. This energy may manifest itself in light particles (photons) or in other particles and other antiparticles.

The reverse is also possible: a particle-antiparticle pair may be created by existing energy, for example, in the form of kinetic energy.

One of the most important principles of quantum theory is that despite the fact that the law of conservation of energy is true in the long term, in the short term the law can sometimes be violated, and it is possible to obtain a short-term "energy loan." The bigger the loan, the more rapidly it must be "repaid." This is one of the outcomes of the "uncertainty principle," a principle formulated by Heisenberg all the way back in 1924.

According to this principle, scientists predicted that the electron inside a hydrogen atom would occasionally transform into an electron combined with an electron-positron pair, who would then destroy one another, thus leaving the atom with only the original electron. And so it was that measurements made in 1947 by Willis Lamb and Robert Retherford discovered a minute difference between results of the Dirac equation and actual measurement results. The difference is known as the "Lamb shift." The Lamb shift can be interpreted due to an electron-positron pair that exists inside the hydrogen atom for about one-millionth of the time in addition to the atom's ordinary electron.

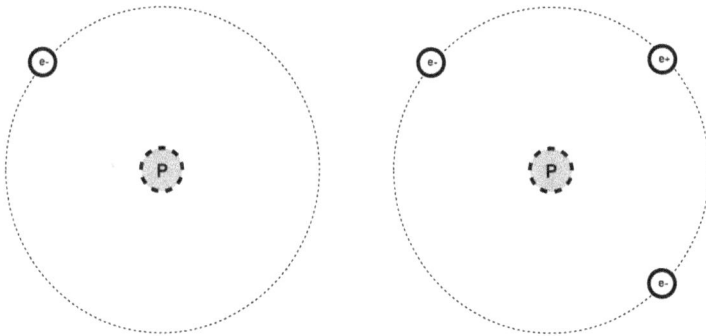

Figure 15. The two states of a hydrogen atom. Left: common atomic state; right: a hydrogen atom with an electron-positron pair.

We are going to use this principle later on in order to understand the structure of protons and neutrons, which contain the other pairs at a much higher frequency. In fact, the proton actually contains an additional quark-antiquark pair in about half of all proton states.

Hadrons

One of the most fundamental questions in studying the structure of protons and neutrons was whether these constitute elementary particles, or whether they are composed of several smaller particles that are bonded to each other. It was only in the 1960s that scientists finally determined that protons and neutrons are indeed *not* elementary particles, although evidence indicating this conclusion has been present since the 1930s. In the meantime, scientists have gathered data on the structure of protons by bombarding protons with photons or electron beams, and by classifying the particles generated as a result of such bombardment. Protons and neutrons are very similar particles, and the use of protons rather than neutrons is easier because free protons are plentifully available in nature (in the form of hydrogen atoms), whereas neutrons barely exist in free form. Furthermore, protons can be accelerated as they possess an electric charge, whereas it is extremely difficult to accelerate a neutron.

When a proton is bombarded and it absorbs energy from the bombarding particle, several outcomes are possible. One, the proton collides with the bombarding particle by way of an "elastic collision," similar to the way billiard balls collide with each other. Under this scenario, the proton will "fly" in a certain direction, whereas the bombarding particle will fly in another direction. Such a collision, as is approximately the case for billiard balls, is in accord with the laws of conservation of energy and momentum.

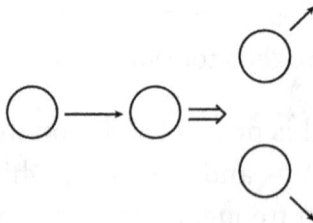

Figure 16. Elastic collision between two particles.
Similar to two billiard balls colliding.

Other possible collisions are known as inelastic collisions. Such collisions result in the formation of new particles. How does this happen? The energy generated by the collision is transformed into a particle-antiparticle pair—namely, the exact opposite of the mutual destruction that occurs when a particle and antiparticle collide. This pair can then produce two novel particles, or a single particle that is related to both, or other particles formed of other combinations of the constituent elements of all particles that are generated after the collision takes place.

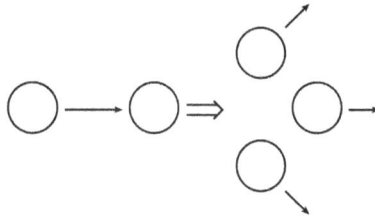

Figure 17. Inelastic collision resulting in novel particles.

Today we know that the proton contains three smaller particles known as quarks, which together comprise about half of the proton's total mass. These quarks can be found inside the proton when it is in its ground state—namely, when it possesses the lowest amount of energy possible. When protons are bombarded, they are sometimes "excited," thus shifting them to a higher energy level. Such an "excited" proton swiftly returns to its original state or to a neutron, usually by emitting particle-antiparticle pairs, thus reducing their energy level to their original state.

It follows then that there are two types of particles that are formed by inelastic collision: "excited" protons known as "baryons," and other particles that are created by particle-antiparticle pairs. In most cases, baryons immediately emit bound particle-antiparticle pairs, and then return to being a stable baryon in the blink of an eye—namely, a proton or neutron. A free neutron is not a stable particle, but it has a very long mean half-life of about fifteen minutes, which is why it can still be regarded as a stable particle.

And just what are the particles that can be formed by their particle-antiparticle pairs? One such particle is known as a "meson." A meson is composed of a quark and an antiquark (not necessarily of the same type). Other pairs can also form, such as electron-positron pairs, muon-antimuon pairs, an electron (or a muon) and its antineutrino (or vice versa), or even baryon-antibaryon pairs. We will address these particle types later on.

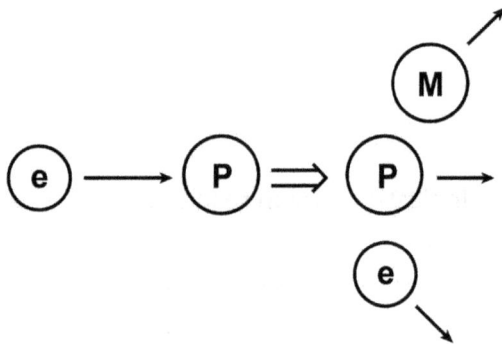

Figure 18. An example of an inelastic collision between an electron and a proton. The proton may become "excited" and transforminto another baryon that immediately disintegrates into a proton and a meson.

Meson particles are also short-lived, and they immediately transform into other particles. The most stable mesons are known as pions (π^+ and π^-) and kaons (K^+ and K^-), which dissolve after a hundredth of a millionth of a second into multiple particles. In particle physics terms, that is actually a very long time. What's important to the current discussion is to understand that we can only see the relatively stable particles detected by our sensors. Such stable particles include photons, electrons, protons, neutrons, pions, kaons, muons, and their respective antiparticles. The measurement of these products enables us to study the properties of the original particle before it disintegrates, even if the original particle had a very short lifetime.

Between the late 1940s and the 1960s, this field of research resulted in the discovery of dozens of new particles. The efforts to understand the significance of these particles, the manner by which they are formed, and the structure of protons and neutrons has resulted in the discovery of quarks.

Gell-Mann, Zweig, and Quarks

In 1964, Murray Gell-Mann and George Zweig separately published what would later become the Quark model, each man publishing a slightly different version. Here I only mention the topics most important for our purposes.

According to Gell-Mann, the proton is composed of three sub-particles, which he named quarks: two u (up) quarks and one d (down) quark. The u quark has an electric charge of ⅔, and the d quark has an electric charge of -⅓. The neutron, however, has two d quarks and one u quark.

In addition, once a proton is bombarded, another kind of quark may also form, known as s (strange), that possess an electric charge of -⅓. This s quark rapidly transforms into either a u or d quark. Many years later, it turned out that there were three other types of quarks, marked by the letters c (charm), b (bottom), and t (top).

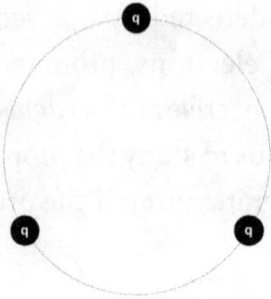

Figure 19. A baryon according to the Quark model contains three bound quarks. Protons and neutrons are baryons.

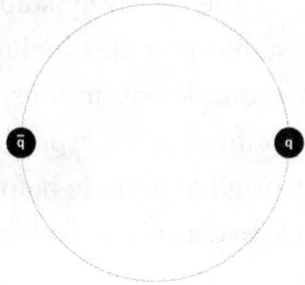

Figure 20. A meson according to the Quark model contains a quark bound to an antiquark.

Gell-Mann and Zweig's ideas allow us to understand the deluge of new particles discovered as combinations of these three *u, d,* and *s* quarks. Baryons are composed of three of these quarks, and mesons are composed of a quark and an antiquark. Each possible combination has several combinations of its own, just as a hydrogen atom can be in its ground state or at higher energy levels. At that point in time there was still no quark-dependant computational model available that could show how the energy levels of baryons and mesons are achieved.

An examination of Gell-Mann and Zweig's articles would indicate that there is only one possible way to describe a baryon. Baryons, according to Gell-Mann and Zweig, are made up of three quarks, and no consideration is given to the possibility that they might contain other massive particles.[44],[45]

---

44   Murray Gell-Mann, *A schematic model of baryons and mesons,* Phys. Lett. **8** (1964) 214–215. *"Baryons can now be constructed from quarks by using the combinations (qqq), (qqqqq-bar), etc."*

45   George Zweig, *An SU3 model for strong interaction symmetry and its breaking.* January 17, 1964. *"Both mesons and baryons are constructed from three fundamental particles called aces... Each ace carries baryon number 1/3 and fractionally charged."*

At that time scientists were yet unaware of the fact that half of the proton's mass is not ascribed to quarks (or to the other quark-anti-quark pairs). Therefore, from a historical point of view, Gell-Mann and Zweig's approach seemed reasonable. Today, however, we can say that the possible models describing the structure of baryons are as follows:

A. Three quarks attracting each other.

B. Three quarks, some attracting and some repelling one another.

C. The quarks are attracting one another. There are three quarks in the outer shell and additional quarks in inner shells (similar to the attraction between protons and neutrons inside the atomic nucleus).

D. A core attracts all three quarks, while the quarks themselves repel one another (similar to an atom with a single shell).

E. A core attracts several quark shells. There are three quarks in the outer shell. The quarks repel one another (similar to the shells of larger atoms).

In each of the C, D, and E baryon models all baryons have the same internal structure, and they differ only in their three outer-most quarks.

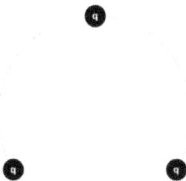

Figure 21. A and B models. The quarks (or some quarks) attract one another. There are no other massive particles.

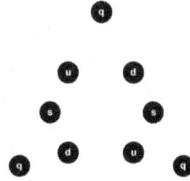

Figure 22. C model. The quarks attract one another, and there are inner shells similar to those of the atomic nucleus.

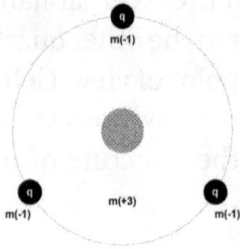

Figure 23. D model. The core possesses a positive strong charge and attracts the negatively strong charged quarks. The quarks repel one another.

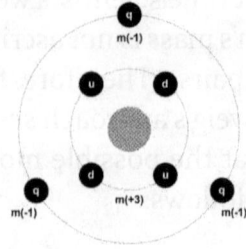

Figure 24. E model. The core possesses a positive strong charge and attracts the negatively strong charged quarks. The quarks repel one another. At least one closed shell of quarks exists.

In the 1960s only models A and B were examined, while the others were disregarded. The B model was proposed by Schwinger,[46] who combined it with the concept of magnetic monopoles (Dirac had a similar concept that addressed the same issue). This idea did not gain any support, probably because its implications would be that neutrons should have a substantial electric dipole moment, and that would be at odds with what we know. That left the A model as the only viable option.

In 1974, physicists were astounded[47] to realize that half of the proton's mass is not carried by the quarks. Paradoxically, instead of discrediting or at least doubting the veracity of the A model

---

46   Julian Schwinger, *A magnetic model of matter,* Science Vol **165** (1969).

47   J. T. Londergan, *Nuclear resonances and quark structure,* International Journal of Modern Physics E **18** 1135 (2009). *"A major surprise occurred with the quantitative understanding of the distribution of the proton momentum."*

and reconsidering the suitability of the C, D, and E models, these models were never brought up for debate.

The Crisis

Even after the quarks concept was published in 1964, Gell-Mann was not convinced that those were genuine particles.[48] In the late 1960s James Bjorken and Richard Feynman devised the necessary mathematical tools needed for calculating the behavior of quarks. And so it was that the experiments conducted with the largest particle accelerator at the time, in the Stanford Linear Accelerator Center (SLAC), corroborated Feynman and Bjorken's predictions, and showed that quarks too were elementary particles with a spin of ½, and therefore quarks had to comply with the Pauli principle.

However, the Δ^{++} baryon was already known to science when the quark concept was first published. If the quark concept is correct, it would then follow that this particle is composed of three u quarks. The Δ^{++} baryon's spin was known to be 3/2 and its parity is even.

For some reason physicists assumed, and many of them still assume, that the three quarks of the Δ^{++} baryon are s-waves at the lowest energy level—namely, situated in the 1s shell. Therefore, since the s-wave does not possess orbital angular momentum, the sum of these quarks' spin is equal to the total spin of this baryon. It follows, then, that in their view a Δ^{++} baryon contains three u quarks of the same quantum state.

---

48   Murray Gell-Mann, *A schematic model of baryons and mesons*, Phys. Lett. 8 (1964) 214-215. *"It is fun to speculate about the way quarks would behave if they were finite particles of finite mass..."*

And here we arrive at a contradiction to Pauli's principle. This is why physicists in the 1960s decided that physics at the time was unable to describe what was happening inside the proton,[49] and that was the reason for the invention of QCD, a theory that rests on assumptions hitherto unknown to physics.

Let's take another look at the A, C, D, and E models, and see if we can find one among them that can describe the Δ^{++} particle.

Even the A model, which results in a contradiction, can explain the nature of this particle if we remember that, according to multiconfiguration theory, which accurately describes a particle made up of three or more sub-particles, such a particle would always constitute a mixture, and it cannot consist solely of s-wave particles. That is the same conclusion we arrived at in the previous chapter on spin.

As for the D model, it too can describe the Δ^{++} particle with even greater ease. Considering the fact that it postulates a baryon composed of four sub-particles (a core and three quarks), it therefore follows that we would need a greater number of effective configurations in order to describe the Δ^{++} particle.

The C and E models depict a Δ^{++} even without understanding the configuration concept. They both describe a baryon with closed shells, and so there is no problem with assuming that the three outermost quarks occupy a p-wave, or any other wave that isn't an s-wave. This is why the quarks possess angular momentum. The E model even explains why the Δ^{++} particle's quarks all prefer to be

---

49   F. Halzen and A. D. Martin, Quarks and Leptons (Wiley, New York, 1984). *"When implementing the quark scheme, however, one runs into a trouble... The uuu configuration correctly matches the properties of Δ++ baryon... Its spin 3/2 is obtained by combining the three identical u quarks in their ground state... Such a state is of course forbidden..."*

in the same direction, in a manner akin to Hund's rule of atomic physics.

Another assumption made by physicists was that the proton's quarks are all s-wave quarks. As we will see later, it is increasingly accepted that the proton's quarks possess orbital angular momentum—namely, they are not solely restricted to s-waves.

The D and E models, according to which the strong force is similar in nature to electric force, also explain why baryons have exactly three quarks. According to these models, the baryon's core (D model), or the core added by the internal shells (E model), have a strong charge of +3 and a total electric charge of zero. Each quark has a strong charge of -1, and each antiquark has a strong charge of +1. Considering the fact that the baryon must be neutral in terms of its strong charge (like an unionized atom), it must have three quarks in its outer shell.

In summary, the QCD theory, which is founded on innovative and fantastic ideas, was created in order to explain a "crisis" that is really no crisis at all.

In the cave allegory I mentioned, after pools were discovered by scientists, they claimed that these pools did not contain any water. One would think that I was just exaggerating when I portrayed the scientists in a ridiculous light, right? Well, let's see if the corresponding story from real-life particle physics is no less improbable:

- In the 1960s scientists thought that there were only three quarks inside the proton and the other baryons and that these particles had no additional quarks.

- Because protons and baryons only had three quarks, these were situated in their innermost 1s shell, which didn't have enough room for three quarks of the same type whose spins were aligned in the same direction.

- Therefore, in order to abide by the Pauli principle and explain the properties of Δ^{++}, they had to come up with QCD, a theory based on ideas hitherto unseen in physics.

- But when it was discovered that there exists a substantial mass of a non-quark matter at the proton, scientists claimed that QCD was nevertheless true and continued to ignore models C, D, and E.

So which one sounds more unlikely in the readers' opinions, the cave allegory or the true story of particle physicists?

Following certain discoveries made in the 1980s, many scientists acknowledged the fact that the proton's exterior quarks do not behave as expected of quarks situated in the innermost 1s shell. These conflicting data are known as the proton spin crisis, and we will discuss them further later. And yet, mainstream physicists are not willing to take the risk and see whether the flaws found in QCD's premises justify a reexamination of the theory.

The Case of Meredith Kercher and Amanda Knox

In societies that subscribe to a common religion, there is almost no event imaginable that might convince them that they are wrong. When the Lubavitcher Rebbe died, his many followers who did not believe that such a thing might happen to their revered Rebbe should have admitted they were wrong. That, however, was not the case. On the contrary, since his death they been even busier trying to convince everyone that the Rebbe is not dead, and that he is about to come back. Instead of shaking their faith, his death made it even stronger.

Such behavior is characteristic of any human society. On many occasions the wisest experts in their fields would present the world with a theory, and, after evidence that contradicts it is found, they simply adjust it instead of replacing it, and succeed in convincing the public that it is still true. This is a natural phenomenon that takes place almost without anyone noticing, and it applies to every aspect of human life. When it does happen, the public may place blind faith in experts, thereby abandoning their critical reasoning abilities altogether. In order to illustrate this phenomenon I present a completely unrelated example, and I apologize if anyone is offended by my reference to this tragedy.

In November 2007 an appalling murder took place at a student's apartment in Italy. Evidence found at the scene indicates that it involved breaking and entering, rape, and murder. Using DNA samples taken from the body and the apartment where the student Meredith Kercher was murdered, and by comparing the fingerprints found at the scene, the police were quickly able to locate Rudy Guede, a young man who had been kicked out of his foster family's home, where he had lived until several months before the murder. Guede had been arrested several times for breaking and entering and was an active drug dealer. He also fled to Germany immediately after the murder. He was eventually convicted of murder in 2008 and sentenced to thirty years in prison.

So that's it, then, right? Not quite. As usual, I tend to present the facts not necessarily in chronological order. During the first few days after the murder the police suspected Kercher's roommate, the student Amanda Knox. Italian investigators had conjured up a narrative whereby Knox and her boyfriend were, in fact, the ones who murdered Meredith. Several days later the police investigator announced that Amanda Knox was the murderer, and that the case was closed. When it turned out that the murderer was actually Rudy Guede, investigators refused to let go of their original theory and claimed that both the random criminal *and* Knox and her boyfriend were involved in the murder.[50]

---

50 en.wikipedia.org/wiki/Murder_of_Meredith_Kercher.

The story was widely covered by the media, which, for some reason, took the side of the investigators. Italy has a jury-based justice system, and Knox and her boyfriend were convicted of murder in 2009, but they were then acquitted by the Italian supreme court in 2011.

Whoever wants to learn more about the subject can take a look at the Wikipedia article and examine the arguments presented by the judges at the time for convicting Knox and her boyfriend.

As far as human behavior is concerned, this story is actually very similar to the behavior associated with the birth of QCD, as the following table shows:

Table 2. A comparison of human behaviors
in two cases that require solving a mystery.

QCD	The Murder of Meredith Kercher
The investigators had uncovered a contradiction in their theories, and were facing a seemingly difficult problem.	The investigators were facing a difficult problem—namely, identifying the murderer.
The investigators created a theory that looked fantastical, but at the time there was no other suitable theory in existence.	The investigators fabricated a fantastical theory because they did not have any viable suspect.
When it turned out that protons contain an additional mass not carried by quarks, the investigators made no effort to accept the necessary solution.	When it turned out that the case had a simple explanation, the investigators refused to let go of their original theory.

QCD	The Murder of Meredith Kercher
The public believed the investigators as it was unable to examine the underlying logic of their claims.	The public believed the investigators and didn't bother to study the details of the case or the "evidence" on which the conviction was based.
Once a certain theory is accepted, it is very hard to let it go.	
When new facts that fail to match the original theory are found, the natural response is to simply adjust the theory. It is likely that such adjustments are almost always based on the belief that the theory is correct.	
A large group of intelligent people may subscribe to a theory that, to almost every objective viewer capable of critical thinking, would seem at odds with reality.	

Remember the chronological evidence test? Try to imagine that Guede had been discovered on the very first day, along with all the evidence he had left at the crime scene. Now, try to imagine that during his first interrogation he hadn't mentioned the presence of Knox and her boyfriend at the apartment (as he did in real life, when he mentioned them only after having learned that the two were suspected of the crime). Would one as an investigator choose to interrogate them, then? Would one ascribe any significance to what was later described as "strange behavior" indicating that Knox and her boyfriend were involved in that terrible crime?

Let's get back to our present discussion. Is QCD at odds with reality as well? The fact that QCD is founded on a flawed premise does not discredit it entirely, although it does weaken it to a considerable extent. Indeed, there is only a very small chance that a theory that rests on countless ideas hitherto unknown to physics, and that is based on erroneous premises, will eventually turn out to be correct, but it is nevertheless possible.

In the following sections we will see how QCD and the C, D, and E models account for the known experimental data obtained so far.

FORCES INSIDE THE ATOMIC NUCLEUS

Would You Like to Solve a Puzzle?

The atomic nucleus is made of nucleons—protons and neutrons. What's keeping them together, and what's keeping them from separating? What are the forces that operate between them?

These questions have been debated for more than eighty years, and they still lack a consistent explanation. In this chapter I will describe several nuclear phenomena, most of which constitute a scientific mystery, and, if readers pay close attention, they may be able to solve that mystery themselves, the same mystery that has been in the forefront of scientific thought for so many years.

Am I going overboard? Maybe. We'll know in just a few pages.

Some Basic Concepts

The atom contains a nucleus surrounded by electron shells. The electrons have a negative electric charge, and each electron repels the others. On the other hand, the nucleus, which has a positive charge, attracts the electrons. This force is known as electromagnetic force (or simply electric force).

The force that keeps the atoms of a noble gas together inside a drop of liquid is known as van der Waals force, named after a Dutch physicist. Van der Waals forces are much weaker than electromagnetic forces. Take a look at the very simplistic illustration here.

The figure shows the nucleus and the atoms as perfect spheres, rather than wave functions, and the nucleus is significantly larger than it really is. Van der Waals forces also operate between molecules, but this fact is irrelevant to this discussion.

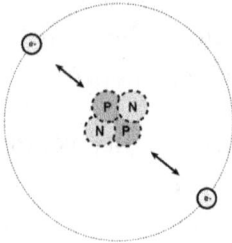

Figure 25. The electromagnetic force operating between the nucleus of a helium atom and its electrons.

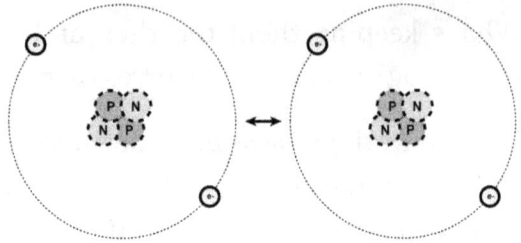

Figure 26. Van der Waals force operating between two helium atoms. See explanation below.

In an experiment conducted with the Stanford Linear Accelerator in the late sixties and early seventies, scientists revealed that each nucleon (proton or neutron) had three quarks situated in the outer shell. A significant result of the experiment showed that a substantial portion of the proton's mass was not carried by the quarks.[51], [52] According to today's predominant theory, QCD, the remaining mass is carried by particles known as gluons. QCD argues that these particles are the ones that glue quarks together. At this point we will ignore this argument and leave the debate about gluons open.

---

51 H. Frauenfelder and E. M. Henley, *Subatomic Physics*, (Prentice Hall, Englewood Cliffs 1991) p. 153.

52 D. H. Perkins, *Introduction to High Energy Physics*, (Addison-Wesley, Menlo Park, CA 1987) p. 282.

The force that keeps quarks together is known as the strong force (or strong interactions).

The force that keeps nucleons inside the atomic nucleus is known as the nuclear force (or strong nuclear force).

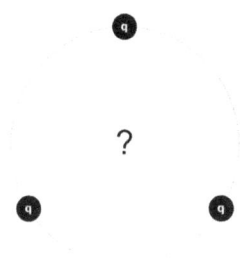

Figure 27. The strong force keeps quarks inside nucleons.

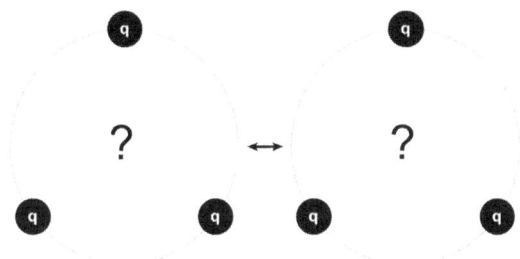

Figure 28. The nuclear force keeps nucleons inside the atomic nucleus.

Van der Waals and nuclear forces share several interesting properties.

Residual Forces that Cancel Out at a Distance

Two fundamental physical forces, the electromagnetic and gravitational forces, operate between two objects, and the strength of force decreases gradually with the distance between the objects.

In contrast, van der Waals forces that operate between noble atoms work differently: when the atoms grow apart, the force is cancelled out completely. It is active only when the atoms are close enough to each other.

How does this happen?

A noble gas atom looks neutral when one measures it from distance—namely, the atom does not apply electromagnetic forces on a distant electric charge. The reason for that is that the negative charges of electrons and the positive charge of the nucleus cancel each other out. This is known as the screening effect.

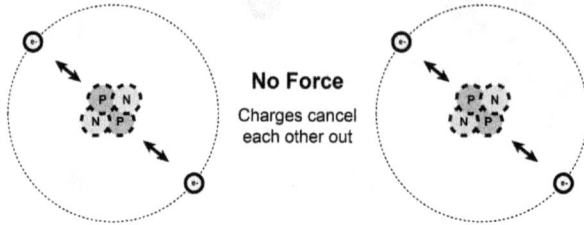

Figure 29. A simple illustration of the screening effect between two helium atoms that contain two electrons each. The nuclear and electron charges cancel each other out whenever the atoms are distant from one another.

When the atoms are close, however, the electrons in the outer shell of one atom can "feel" the electric charge of the other, thus inducing changes in the distribution of the shell's electric charge. As a result, the attractive force between the nucleus of one atom and the electrons of the other becomes stronger than the repulsive force between the electrons of the two atoms. In this fashion, the atoms are bound together at low temperatures and are thus liquefied.

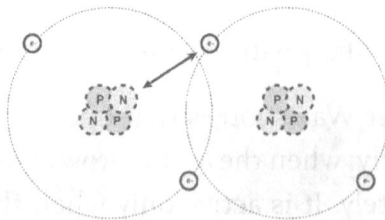

Figure 30. Van der Waals forces operating between two helium atoms. The electrons of one atom are attracted to the nucleus of the other.

Similarly, the nuclear force operating between nucleons is very strong, but it is much weaker than the strong force keeping the quarks together inside the nucleons. In addition, when nucleons grow apart from each other, this force ceases to affect them completely. In this regard, it is very similar to how van der Waals forces behave.

Both nuclear and van der Waals forces are known in the literature as residual forces. They are much weaker than the forces from which they stem: van der Waals forces are much weaker than the electromagnetic forces that operates between the positive nucleus of the atom and its electrons, and the nuclear force is much weaker than the strong force that operates between quarks.

Incompressible

One of the known properties of liquids is that they are incompressible; namely, their volume is almost unaffected by pressure or temperature changes. In fact, the specific volume (the volume of a unit of mass) of liquids remains almost constant, since at first the atoms grow closer together as a result of van der Waals forces, but at very small distances a very strong repulsive force arises due to the Pauli principle. An identical phenomenon occurs inside large nuclei. The nucleons inside the nucleus have a fixed density, and are never compressed.[53] QCD is unable to explain this phenom-

---

53   S. S. M. Wong, *Introductory Nuclear Physics*, (Wiley, New York 1998) p. 139.

enon, and, as we will see later, the developers of QCD themselves have admitted that it contradicts the theory.[54]

The Relationship between Distance and Force

The graph in Figure 31 shows the relationship between the force[55] that operates between two noble argon gas atoms as a function of distance. The sharp decline seen on the left side of the graph depicts the greater repulsive forces that exist when atoms are so close to each other so as to create a substantial overlap between their electrons. The graph goes below zero, and when it goes up one can see an attractive force which becomes steadily stronger, steadily weaker, and then goes back to zero. All of this occurs as atoms grow apart. The repulsion stems, as we can already tell, from the Pauli principle, the attraction from van der Waals forces, and the ultimate cancellation of the force is caused by the screening effect brought about whenever the atoms are distant enough from each other, and which causes van der Waals forces to cancel out.

A similar interaction occurs between two nucleons, as the graph in Figure 32 demonstrates. This amazing similarity is also left unexplained.[56]

---

54   Frank Wilczek, *Hard-core revelations*, Nature, Vol. **445** 156 (2007).

55   To put it in more accurate terms, this is the potential from which the force is derived.

56   S. S. M. Wong, *Introductory Nuclear Physics*, (Wiley, New York, 1998). p. 102.

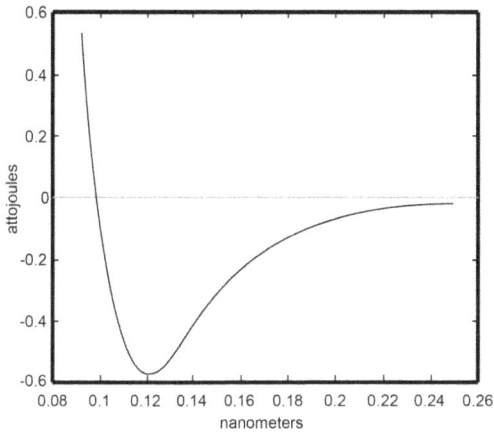

Figure 31. Potential as a function of the distance between two argon atoms.[57]

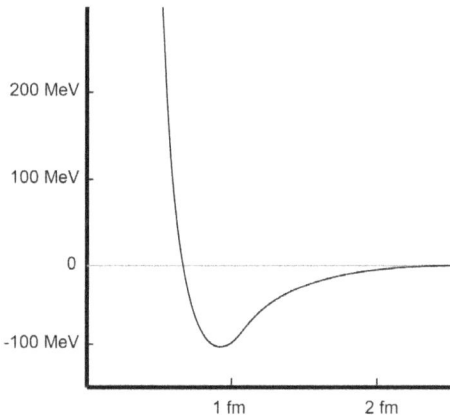

Figure 32. Potential as a function of the distance between two nucleons.[58]

---

57 For a similar graph, see H. Haken and H. C. Wolf, Molecular Physics and Elements of Quantum Chemistry, (Springer, Berlin 1995). p. 15.

58 S. S. M. Wong, Introductory Nuclear Physics, (Wiley, New York, 1998). p. 98.

The Volume of Atoms in Liquids

It turns out that the volume of electron orbits in a free noble atom—namely, in a gas—is smaller than the volume of electrons in the same atom when in liquid form.[59] How does this happen? A liquid state is characterized by the operation of van der Waals forces, which involve the electrons of one atom being attracted to the nucleus of its neighboring atoms. Consequently, in liquid the orbits of electrons inside noble atoms expand, and in fact there exists a small partial overlap between the electrons of one atom and those of another.

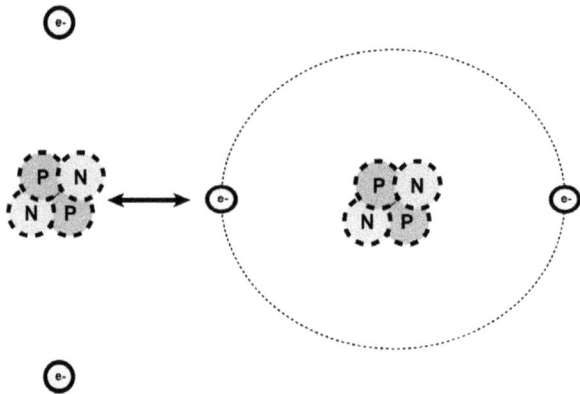

Figure 33. The volume of a helium atom in a liquid state is larger because of the attraction of its electrons to the nuclei of neighboring atoms.

A similar phenomenon was discovered in nucleons. In 1983 it was found that the volume of quarks in nucleons is larger when they

---

59   J. B. Pendry, *The electronic structure of liquids*, J. Phys. C, **13**, 3357 (1980).

are situated in a larger nucleus, as opposed to their volume when situated inside a smaller nucleus that only contains a proton and neutron.[60]

This effect is known as the EMC effect, and it contradicts every single prediction derived from QCD, and remains unexplained to this day.[61]

The Nuclear Tensor Force

According to the theory of electricity, a closed circulation of electric current or an atom that has spin generates an axial magnetic dipole—namely, a magnet with two poles. When two such magnets are situated next to each other, they exert forces on each other. The direction of these forces is usually not oriented on the straight line that connects them. Anyone who has ever played with magnets must have noticed this tensor force phenomenon.

Some atoms, such as the hydrogen atom, have spin, and their electrons produce a magnetic dipole. When two hydrogen atoms get closer to each other, in principle they exert tensor forces on one another. There is another phenomenon, presently irrelevant to this discussion, whereby two atoms form a hydrogen molecule whenever they become close to each other.

---

60   J.J. Aubert *et al.*, Phys. Lett. **123B**, 275 (1983). The graph on p. 277 shows that the x dependence of the structure function of iron is narrower than that of the deuteron. It follows that iron's quarks are enclosed in a larger spatial volume than that of the deuteron.

61   J. Arrington *et al.*, *New Measurements of the EMC Effect in Few-Body Nuclei*, J. Phys. Conference Series **69**, 012024 (2007). *"So while the experimental signature is clear, the interpretation of this effect is, at present, ambiguous."*

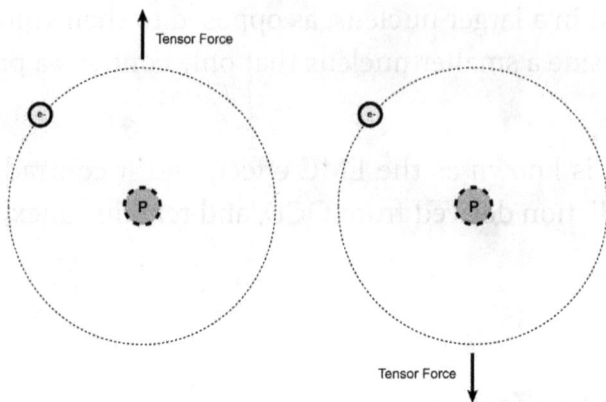

Figure 34. Atoms with spin create a magnetic dipole
and exert tensor forces on each other.

It appears that tensor forces also operate between nucleons, and in a more effective fashion. Protons and neutrons have spin, and they exert tensor forces on each other. Even though this phenomenon has been known to science since 1939,[62] the question of what creates this force remains unanswered.

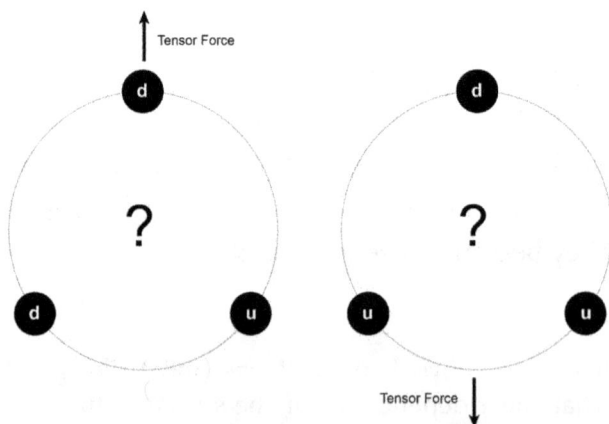

Figure 35. Nucleons exerting tensor forces on each other.

---

62   J. Schwinger, *On the neutron-proton interaction*, Phys. Rev. **55**, 235 (1939).

Let's Compare

Table 3. A comparison between electromagnetic and strong forces.

	Electromagnetic Force and Electrons	Strong Force and Quarks
1	Keeps electrons inside the atom. Relatively strong.	Keeps quarks inside the nucleon. Relatively strong.
2	Keeps atoms inside a drop of liquid due to a much weaker residual force (van der Waals).	Keeps nucleons inside the nucleus due to a much weaker residual force (nuclear force).
3	The residual force rapidly cancels out when the atoms are distant from one another.	The residual force rapidly cancels out when the nucleons are distant from one another.
4	Liquids have an almost constant density.	The nucleus has an almost constant density.*
5	The volume of electrons inside a liquid noble atom is larger than it is inside a free atom.	The volume of nucleon quarks inside a heavier nucleus is larger than their volume inside a deuteron (EMC effect).*
6	The graph showing the potential between the atoms as a function of distance is similar to a ski jump.	The graph showing the potential between the nucleons as a function of distance is similar to a ski jump.*
7	Atoms with spin exert tensor force on one another.	Nucleons have spin and they exert tensor force on one another.*

*Unexplained phenomena

It turns out that the properties of noble gases are remarkably similar to those of nucleons when one replaces an atomic nucleus with
a drop of liquid, nucleons with atoms, and quarks with electrons.
Most of these phenomena have remained unexplained since they
do not correspond with today's leading theory, QCD.

Is nature trying to tell us something?

Do you, dear readers, have a solution for this mystery?

Here I recommend that readers take a short break and invent a
structure for protons and neutrons that is compatible with the scientific findings that were presented in this chapter. Who knows,
maybe I wasn't exaggerating a few pages earlier when I expressed
hope that you would solve the problem by yourselves.

The Strange History of Nuclear Force

Research into the nuclear force that keeps nucleons together inside the atomic nucleus was already taking place in the 1920s. However, even today scientists give different answers regarding the origin of this force, while others will simply admit that the question of what causes the nuclear force remains open. How did that happen?

The Liquid Drop Model

The similarity between atoms inside a drop of liquid and the nucleons inside the atomic nucleus has been known to science for over eighty years. In 1928, the young Russian physicist George Gamow (born Georgiy Antonovich Gamov) described the disintegration of the atomic nucleus, a process known as alpha decay, by likening the atomic nucleus and the forces that operate in it as a drop of liquid.[63], [64]

---

63 G. Gamow , *Zur Quantentheorie des Atomkernes*, Z. Physik **51**, 204 (1928).

64 Roger H. Stuewer, "*The Origin of the Liquid-Drop Model and the Interpretation of Nuclear Fission,*" Perspectives on Science, **2** 1 (1994). p. 76–129.

Gamow was an extraordinary scientist. He left the Soviet Union in 1933 and settled in the United States in 1934. He later presented an explanation for the process of beta decay and turned to the field of cosmology. Together with Ralph Alpher he authored two ground-breaking papers that proved to substantially support the Big Bang theory. A versatile polymath, he helped DNA researchers explain the way DNA works by using RNA molecules.[65]

He authored several classic popular science books that are still in print today,[66] more than fifty years after they were written. Gamow was described by those who knew him as an inspiring figure.

In 1928 Gamow applied the liquid drop concept a long time before scientists knew the relationship between potential and the distance between nucleons (as shown in the preceding chapter). By then, the existence of neutrons was merely speculated, and the final proof of their existence was discovered only in 1932. At the time scientists did not yet know that the proton is not an elementary particle, that it possessed a quark shell, or that there exists a substantial amount of mass inside the proton that is not carried by quarks. And so, even though Gamow was an extraordinary individual, it's hard to imagine that he considered the internal structure of the proton to be similar to that of the atom as early as in 1928.

---

65   American Treasures of the Library of Congress, *DNA: An "Amateur" Makes a Real Contribution.* "*In 1954, George Gamow made what has been called "perhaps the last example of amateurism in scientific work on a grand scale." Less than a year after James Watson and Francis Crick discovered the molecular structure of DNA, Gamow, a professional physicist and amateur biologist, proposed the first definite coding scheme for DNA.*"

66   George Gamow, *One Two Three...Infinity: Facts and Speculations of Science*, Viking Press, 1947. Reprinted by Dover Publications, ISBN: 978-0-486-25664-1.

Additional Successes of the Liquid Drop Model

In addition to the successful application described earlier, the liquid drop model was successfully utilized in order to explain a variety of phenomena associated with atomic nuclei. Nuclear physicists in the 1930s were more interested in studying the atomic nucleus rather than the internal structure of protons. For this reason, it's likely that they used the liquid drop concept without possessing a great deal of interest in the reason for the similarities between atomic nuclei and liquid drops.

One of the more notable triumphs of the liquid drop model is exemplified in a formula developed by German physicists Hans Bethe and Carl Friedrich von Weizsäcker, designed to predict the mass of atomic nuclei. The forces that operate inside atomic nuclei include more than just the attractive force similar to the forces operating inside a drop of liquid. Bethe and Weizsäcker incorporated all relevant forces into their formula and were thus able to rather accurately predict the mass of atomic nuclei.[67] The formula is known as the Bethe-Weizsäcker Semi-Empirical Mass Formula or simply the Bethe-Weizsäcker Formula. Incidentally, the younger Weizsäcker brother, Richard von Weizsäcker, later became president of Germany.

Another success of the liquid drop model is its formula for the radius of nuclei. This radius formula has received experimental confirmation and is also used as an element in the Bethe-Weizsäcker semi-empirical mass formula.

Niels Bohr, aided by his young assistant Fritz Klacker, also used the liquid drop model in order to show how an alpha particle (a helium nucleus composed of two protons and two neutrons) or a

67   C.F. Weizsäcker, *Zur Theorie der Kernmassen*, Zeitschrift für Physik **96** (1935) p. 431–458

neutron is "glued" to atomic nuclei when fired at it. According to Bohr and Klacker, the nucleus oscillates like a drop of liquid in a semi-stable state until it stabilizes.

The most known discovery that owes its existence to the liquid drop model is that of nuclear fission. This discovery, made by Otto Hahn, Lise Meitner, and Otto Frisch,[68] eventually led to the development of the nuclear bomb, and later to the design of nuclear reactors used for generating power. Meitner and Frisch used the liquid drop model to show that nuclear fission was similar to the division of a liquid drop, a process that, according to their calculations, should bring about the release of approx. 200 MeV for each uranium atom.[69]

How did it come to pass that despite the colossal success of the liquid drop model, its implications for understanding the structure of nucleons were never considered? We should assume that had a messenger from the future arrived to tell scientists in the 1930s that nucleons are particles made of quarks arranged in their outer shell, and of an additional mass located at their center, the atomic models (D, E) would then have been seriously considered. However, as far as we know, a messenger from the future did not travel back to the 1930s, or, if one did, he or she was not very successful at convincing anyone.

---

68   Lise Meitner and O.R. Frisch, *Disintegration of Uranium by Neutrons: A New Type of Nuclear Reaction*, Nature **143** (1939). p. 239–240.

69   Sime, Ruth Lewin, *Lise Meitner*, California Studies in the History of Science, 1996. p. 246–247. University of California Press. In their paper the two physicists likened nuclear fission to the "essentially classical" division of a liquid drop and estimated that the 200 MeV of energy released was "available from the difference in packing fraction between uranium and the elements in the middle of the periodic system."

In this book we will deviate from the common usage of the phrase "liquid drop model," which conventionally describes only the phenomenological similarity between noble atoms in liquids and nucleons inside the nucleus. We will go beyond that and consider this phrase as one that reflects a structural similarity between noble atoms and nucleons. The implication of this would be, similar to a noble gas atom composed of a nucleus with a positive electric charge and negatively charged electrons, that the model describes protons and neutrons as particles with a positive strong charge at their core and quarks as particles with a negative strong charge that surround the core. In this fashion, the quarks inside the proton play a part that is analogous to the role of electrons in the atom, and the laws of strong forces are analogous to the laws of electromagnetic forces.

Yukawa's Theory

In 1932 James Chadwick empirically confirmed the existence of the neutron. Measurements showed that the mass of protons was almost identical to that of neutrons.

Later that year, Heisenberg published the first explanation[70] of the force that operates between protons and neutrons. Heisenberg described that force by positing a positively charged particle that was transferred from protons to neutrons, thus leading them to exchange their roles. Heisenberg proposed that neutrons become protons when they emit an electron, and protons become neutrons when they absorb that electron. These emitted and absorbed electrons produce the nuclear force that attracts protons to

---

70 W. Heisenberg, *über den Bau der Atomkerne. I*, Zeitschrift für Physik **77** (1932).

neutrons. Even though Heisenberg's theory is incorrect, the idea that there is a "force-carrying" particle began to gain popularity among physicists.

In 1934, Wolfgang Pauli and Victor Weisskopf published an article that demonstrated how the Klein-Gordon equation could describe the wave function of a massive elementary particle that has spin zero.[71] Such a particle was unknown at the time, and Pauli himself thought the idea was at odds with reality.[72] Nevertheless, the equation is still included in today's physics books, despite the fact that even the Standard Model argues that there is no particle in the universe that conforms with this equation. Dirac himself had criticized the equation and decreed it to be erroneous, although scientists at the time attributed his objection to personal motives and assumed that this was his natural response to an equation that sought to compete with the Dirac equation (which describes a massive elementary particle with spin ½).

In the 1930s physicists believed that protons and neutrons were elementary particles, despite the fact that in 1933 Robert Frisch and Otto Stern[73] demonstrated that the proton possesses a magnetic moment that fails to conform to the Dirac equation, and therefore is probably not elementary.

Nevertheless, in 1935 the Japanese physicist Hideki Yukawa published a theory that explains the force that operates between protons and neutrons. Yukawa's theory argues that protons and neutrons are elementary particles, and that there exists another

---

71   W. Pauli, V. Weisskopf, Helv. Phys. Acta 7, 709 (1934).

72   A. I. Miller *Early Quantum Electrodynamics* (University Press, Cambridge, 1994). p. 70.

73   Frisch, R. and Stern, O., *Über die magnetische Ablenkung von Wasserstoffmolekülen und das magnetische Moment des Protons*, Zeitschrift für Physik, **85** (1933).

elementary particle, later called the Yukawa particle, that "carries the force" that operates between nucleons. According to Yukawa, this new particle lacks spin, conforms to the Klein-Gordon equation, and carries the force that operates between protons and neutrons. Yukawa also demonstrated that the mass of this particle ought to be approximately 100 MeV.

In the 1940s scientists discovered the π^0 particle, which appeared to be spin zero. The mass of this particle roughly corresponded to the mass predicted by Yukawa. As a result, Yukawa's theory was recognized as the theory for describing the force that operates between nucleons, and Yukawa was awarded a Nobel Prize for his theory.

Today, however, we know that the π^0 particle is not elementary, that nucleons are not elementary, and that there exists no elementary particle in the universe that conforms to the Klein-Gordon equation.

Wait a minute! If all of Yukawa's assumptions turned out to be false, then how is Yukawa's theory relevant in any way to this discussion?

As it turns out, Yukawa's theory is *still* presented in today's textbooks, simply because no other theory was ever found to replace it!

In the 1960s, following the discovery of quarks, physics professors were asked by their students why they persist in teaching Yukawa's theory despite the fact that it is theoretically unfounded. The professors promised their students that, in the near future and after scientists better understand the forces that operate between quarks, they would uncover the correct theoretical explanation for the force that keeps nucleons inside atomic nuclei—namely, the nuclear force.

We are still waiting for this explanation.

Quantum Chromodynamics

When quarks were discovered in the 1960s, physicists realized that protons and neutrons were not elementary particles, and that they contained three quarks. It was only in 1974 that scientists discovered that approximately half of the particles' mass cannot be attributed to quarks. Quantum chromodynamics, or QCD, was first published in 1972, and rested on two key premises:

- The quarks possess all of the nucleon's mass.

- All proton quarks (and those of other particle, such as Δ^{++}) are pure s-waves.

These two premises led to the sixties crisis described in previous chapters. QCD was founded on these two premises, provided an explanation for the existence of the Δ^{++} particle, and also explained why all baryons have exactly three quarks. Following is a summary of what the theory actually says:

According to QCD, each quark possesses a charge called "color," and there are three possible colors for each quark. For illustration purposes, the colors are known as red, green, or blue.[74] According to QCD, only particles that have an equal amount of colors (ergo, if we use these three colors, such particles would be white) can exist in free form. Antiquarks have anticolors of the same type, and therefore mesons can exist in free form as well.

The colorful charge of QCD extends the "space" inside baryons threefold, without violating the Pauli principle, as in this fashion there could be three quarks with the same quantum state situ-

---

74   These colors are just a verbal way of denoting three types of charges that act together and are entirely unrelated to the colors we see with our eyes.

ated simultaneously inside the baryon, provided each of them has a different color. This explains the existence of the Δ^{++} particle, which has three u quarks and a spin of 3/2.

QCD argues that there exists a type of massless particles (namely, gluons) that are found in baryons and mesons and that carry the color force. Each gluon carries a color and an anticolor of a different type, and they actually transfer the colors from one quark to another and are responsible for the attractive force between quarks.

QCD also presents us with new force equations. The equations are highly elaborate, and to this day they are considered virtually unsolvable. To solve them accurately, even when using the fastest computers available, is impossible, and therefore certain approximation methods based on a technique known as Lattice QCD have been developed. These methods are in some cases able to generate a reasonable estimate.

When it turned out in the mid-1970s that half of the proton's mass cannot be attributed to quarks, physicists were astounded.[75] At that point, however, even though one of the fundamental premises of QCD was found to be wanting, no one ever considered discarding the theory or examining any alternatives. Instead, scientists simply decided that the missing mass was carried by gluons.

A New Language

QCD theory is not a gradual extension of the theories that preceded it—namely, quantum theory and the quantum field theory of

---

75   J. T. Londergan, *Nuclear resonances and quark structure,* International Journal of Modern Physics E **18** 1135 (2009). *"A major surprise occurred with the quantitative understanding of the distribution of the proton momentum."*

electromagnetic forces. Not only is it based on a large number of ideas that have no counterpart in nature, its mathematical framework actually consists of a whole new language, which prevents physicists in other fields from taking part in the academic discussion on strong forces.

To illustrate, I present an excerpt from a letter sent to me by Freeman Dyson in 2011. Dyson is one of the four founders of quantum field theory. Quantum field theory had two "versions," one by Schwinger and Tomonaga and another by Feynman. Dyson proved that the two approaches were equivalent and paved the way for the acceptance of quantum field theory.

Following is the excerpt in question:

"...I gave up doing particle physics about forty years ago and have not tried to understand the more recent work in that field. I never worked with QCD and never studied it in detail...Here I am surrounded by brilliant young people whose language I do not understand..."[76]

If Dyson can't understand the language of QCD, then who are we to even try? Indeed, in this book we would normally make do with the fact that even scientists who adhere to QCD admit its flaws.

The history of QCD is a strange one, but perhaps it nevertheless accurately portrays the forces that operate between nucleons? We will seek to answer this question in the following chapters.

---

76   Freeman Dyson, private letter, 6 August, 2011.

QCD and the Forces between Nucleons

The nuclear forces that operate between nucleons are so utterly incommensurate with QCD that even Frank Wilczek, who was awarded a Nobel Prize for a discovery related to QCD, admitted that as far as QCD is concerned nuclear forces are entirely groundless.[77]

There is an abundance of physical phenomena that contradict QCD theory and that relate to forces between nucleons. Let's elaborate on them.

The Similarity between Nuclear and van der Waals Forces

The attractive forces operating between nucleons, which are similar to van der Waals forces, cannot be explained by QCD. Not only has it been over forty years since QCD was first presented and no one has yet shown how this force fits with the theory, famous physicists even went so far as to state that it fails to provide a correct description of those forces.[78]

---

77 Frank Wilczek, *Hard-core revelations*, Nature, **445** 156 (2007). *"Ironically, from the perspective of QCD, the foundations of nuclear physics appear distinctly unsound."*

78 S. S. M. Wong, *Introductory Nuclear Physics*, (Wiley, New York, 1998). p. 102. *"Currently, the color van der Waals force does not seem to be a correct model for nuclear interaction without modifications."*

The constant density of nucleons inside the atomic nucleus also remains unexplained by QCD. Soon, as we describe a simple nucleus containing exactly two nucleons, we will see how this manifests itself in the most severe fashion.

The EMC effect, which showed that the nucleon's quarks inside large nuclei have a larger volume, includes a contradiction to QCD theory, as I mentioned in the beginning of this book.[79] The effect remains unexplained to this day.[80]

QCD and the Deuteron

The nucleus of heavy hydrogen, or deuteron, contains a proton and a neutron that are "glued" together. The proton contains three quarks (*uud*) and the neutron contains three quarks *(udd)*.

Since QCD's color scheme essentially augments the contents of protons threefold, it follows that protons have ample available space, and the Pauli principle does not prevent the proton and neutron from combining, along with the six quarks that comprise them, into a single particle.

This is explained by Wilczek: he says that QCD tells us that quarks of the proton and neutron are trapped inside "bags," the contents

---

79   J.J. Aubert *et al.*, Phys. Lett. **123B**, 275 (1983). *"The results are in complete disagreement with the calculations...We are not aware of any published detailed prediction presently available which can explain the behaviour of these data."*

80   J. Arrington *et al.*, *New Measurements of the EMC Effect in Few-Body Nuclei*, J. Phys. Conference Series **69**, 012024 (2007). *"So while the experimental signature is clear, the interpretation of this effect is, at present, ambiguous."*

of which should contain an equal amount of each color. He then asks, "But why don't the separate proton and neutron bags in a complex nucleus merge into one common bag?" Wilczek later explains that such a bag would result in a lower energy level and should therefore be preferable to two separate bags. Wilczek expressed his hope in his article that a future breakthrough will eventually resolve the conflict.

Naturally, not only Wilczek is aware of this contradiction, as illustrated by the following correspondence taken from an online forum for physicists. The physicists do not doubt the veracity of QCD, but they are nevertheless aware of the contradiction.[81]

> … In the end it [the deuteron] should come from some fundamental computations in QCD. Maybe it does, but I am not aware of any published work on that. (I hope I am wrong, I will be the first to check and read it, if it exists.) Worse yet, there are no QCD calculations that I know of which explains why there are no six quark color singlet states. Of course, one might consider Deutorium a six quark color singlet, but it does not cut the muster. Because it is not really a six quark color singlet; it is a bound state of two more-or-less spatially separated three-quark color singlets…

The Deuteron and Tensor Forces

The deuteron is the only stable nucleus that contains precisely two nucleons. It contains a proton and a neutron whose spin is in the same direction—namely, the deuteron's total spin is 1.

---

81   www.physicsforums.com/showthread.php?t=370657

Let's discuss for a moment the question of why it is that this is the only stable combination possible. The possibilities at our disposal are:

- Two nucleons with opposite spins: In this scenario, the force that keeps the two nucleons together is the nuclear force—namely, an equivalent of van der Waals forces. It turns out that this force is too weak to "glue" two nucleons together.

- Two protons or two neutrons with parallel spins: This scenario is ruled out by the Pauli principle, which applies to the nucleus (protons and neutrons are fermions).

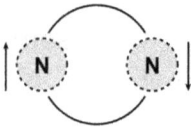

Figure 36. Two neutrons with opposite spins. Van der Waals forces are not strong enough to keep them together.

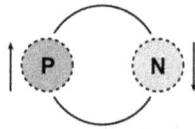

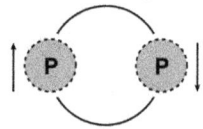

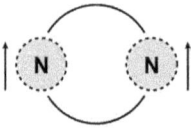

Figure 37. A proton and neutron with opposite spins. Van der Waals forces are not strong enough to keep them together.

Figure 38. Two protons with opposite spins. Van der Waals forces are not strong enough to keep them together.

Figure 39. Two neutrons with parallel spins. Ruled out by the Pauli principle, which applies to fermions (such as neutrons).

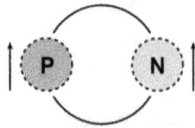

Figure 40. A proton and neutron with parallel spins. "Glued" together by nuclear and tensor forces.

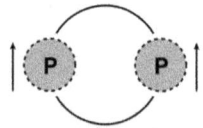

Figure 41. Two protons with parallel spins. Ruled out by the Pauli principle, which applies to fermions (such as protons).

This leaves us with the scenario mentioned in Figure 40: a proton and a neutron with parallel spins. Under such circumstances the Pauli principle is observed, and the van der Waals force is still too weak to keep the two nucleons together by itself. However, here it is aided by an additional force, the nuclear tensor force described in a previous chapter. This force provides significant support for the nuclear van der Waals force, and when combined they manage (barely) to keep the proton and neutron "glued" together.

The origin of nuclear tensor forces remains unknown to this day.

What might bring about the nuclear tensor force? According to conventional physics, tensor forces ought to have origins in a similar phenomenon to magnetic dipoles—namely, a combination of charges that exist in quarks, and in the fact that quarks generate an orbital or self-spin. However, two possibilities present themselves: one is that the electric charges of quarks are the source of this force; and the second is that the strong force operating between quarks results from a strong charge that is similar in some respects to electric charge, and it too can generate an analog of magnetic dipole, which we might call a strong dipole as it is created by a strong rather than electric charge.

Calculations were made to assess the first option, whereby the tensor force stems from the magnetic dipole created by the electric charge of the quarks. This possibility was ruled out for two reasons. First, this force is simply too weak. The nuclear tensor force is immeasurably stronger than the force that could have been generated by the nucleon's magnetic dipole. The other reason was that tensor forces are measured with the opposite sign of the calculated force that is created by the proton and neutron's magnetic dipole.

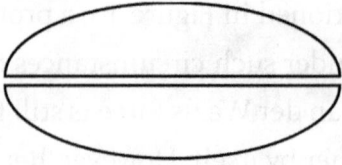

Figure 42. The projected shape of the deuteron if the tensor force is generated by the quarks' electric charge. The deuteron should in this case be similar to a hamburger bun.

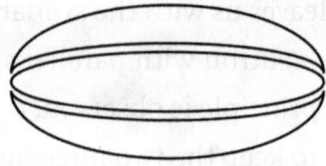

Figure 43. The shape of the deuteron as confirmed by the experiment. The deuteron is similar to a rugby ball.

This leaves us with the other option: tensor force owes its existence to the similarity of strong and electromagnetic forces, and to the strong dipole it generates. However, to this day, despite the fact that we have been aware of tensor forces since 1939,[82] no one has yet demonstrated how QCD might explain their existence.

---

82   J. Schwinger, *On the neutron-proton interaction*, Phys. Rev. **55**, 235 (1939).

Strong CP Problem

A mirror world is a theoretical world where left and right interchange. Under this transformation the theory of electrodynamics would produce the same electromagnetic phenomena found in our world. The implication of this is that left is not preferable to right in such a world and vice versa. In such cases physicists would say that parity (denoted by P) is conserved. Our experience confirms this property of electromagnetic processes.

To all of us this phenomenon appears completely intuitive. What reason would nature have to choose one side over another? Parity is also conserved in the electromagnetic system that I described in the beginning of this book in respect to electrons inside atomic shells.

In the 1950s, physicists were surprised to realize that there are certain particles (which today are called kaons: K^o, K^+, K^-) that decay into two and sometimes three pions. The parity of all pions has been measured, and they are known to be of odd parity. Consequently, when a kaon decays, it does not conserve its parity, and from that it follows that kaon decay behaves differently on one side (say, right) from its behavior on the other (left). Owing to the fact that until that time processes that fail to conserve parity were unknown to science, the phenomenon was considered to be a puzzle known as the τ-θ puzzle.

Today we know that kaon decay is produced by the weak force. It was known at the time as well, and in the mid-fifties Chen-Ning

Yang and Tsung-Dao Lee published a theory whereby the weak force did not conserve parity.

Scientists decided to see whether another form of weak decay, known as beta decay, would prefer one side over another. In a famous experiment conducted in 1956 and known as Wu's experiment (after the Chinese American scientist Chien-Shiung Wu), researchers examined the nuclei of a cobalt atom, which has a magnetic moment, and, by using a strong magnetic field, they "straightened" the cobalt nuclei so that each was facing the same direction. Later they measured whether the beta decay of these cobalt nuclei—which resulted in a neutron decaying into a proton, an electron, and an antineutrino—would prefer one side over the other.

And indeed it did. This result was astonishing despite the fact that it already stemmed from considerations based on the analysis of the kaons' (K^0, K^+, K^-) weak decay, and it was still astonishing because we would intuitively object to the idea that some processes are different when occurring in the mirror world. Today we know that processes that involve weak force may result in parity violation, and we've also identified that electromagnetic processes *do* conserve parity.

Parity and the Strong Force

According to QCD equations, the strong forces do not necessarily conserve parity. QCD equations include a coefficient known as θ, which, when different from zero, means that the strong force will *not* conserve parity. The value of θ ought to be obtained through experimental measurements.

However, according to statistical data gathered to this day, it appears that processes associated with the strong force have no left or right side preference whatsoever, and this statement is accurate to more than ten decimal places. Some QCD proponents view this as no more than a calibration problem. For some reason, by pure chance, θ is exactly equal to zero, despite the fact that according to some Standard Model considerations this coefficient should not be significantly different from 1. Yet others see this as one of the most troubling issues of the Standard Model.[83]

If θ isn't equal to zero, then we should arrive at another asymmetry. In a world where antimatter replaces matter, particles' charges reverse their signs. In this world, the laws of electromagnetism are nevertheless conserved. However, according to QCD equations, this alternate world ought to behave differently if θ is different from zero. This symmetry is known as charge conjugation and is marked by the letter C (hence the paradox's name, the strong CP problem—C for charge conjugation and P for parity).

The paradox seems rather trivial at first. Let's just replace θ with zero and thereby correct QCD's equations! Moreover, it appears that QCD proponents are also divided on the issue, with some claiming that the theory itself conserves parity.[84] But, as we will now see, it is difficult to truly figure out the full meaning of the fact that strong forces conserve parity.

---

83 profmattstrassler.com/articles-and-posts/particle-physics-basics/c-p-t-and-their-combinations/, "But if the strong nuclear force, which holds the neutron together, violates CP, then we'd expect to see an electric dipole moment of 10^{-15} e cm or so. Yet experiment shows that the neutron's electric dipole moment is less than 3×10^{-26} e cm!! **That's over ten thousand million times smaller than expected.** And so the strong nuclear force does not violate CP as much as naively anticipated."

84 Cumrun Vafa and Edward Witten, *Parity conservation in Quantum Chromodynamics*, Physical Review Letters, Vol **53**, 1984.

The Particle That Carries Nuclear Forces

Remember the Yukawa particle? I'm referring to the particle that according to Yukawa's theory ought to carry the nuclear force that keeps protons and neutrons together inside the atomic nucleus. Even though Yukawa's theory argues that protons, neutrons, and the Yukawa particle are all elementary, and today we know that none of them is elementary, plenty of textbooks still exist where this idea is presented as valid, with π^0 playing the role of the Yukawa particle.

These books tell us of "pion clouds" that cover the nucleons and transmit the nuclear force that binds them together. On Wikipedia, for example, there is an amusing illustration showing a single π^0 pion emitting the proton on its way to the neutron or to another proton, and in this fashion carrying the nuclear force.[85]

Such pion emission does require energy, and seemingly contradicts (temporarily) the law of conservation of energy. But such contradiction is permitted according to quantum theory, as short-term energy loans are indeed possible, and the shorter the term of the loan, the bigger the loan can be. Nevertheless, the pion emission idea is unlikely, as the size of the loan required to create a pion is large (135 MeV), whereas the nuclear force is much smaller.

However, if we accept that strong forces preserve parity, and we remember that pions carry odd parity, the whole idea of pion clouds becomes even more groundless. It therefore follows that upon a pion emission by the proton, the proton must also become odd (protons and neutrons are even). In order to do that, it must undergo excitation, which requires a much greater amount of energy,

---

85   en.wikipedia.org/wiki/Nuclear_force.

and therefore the energy loan needed under such circumstances is significantly higher than the amount needed to create a pion. Therefore, the idea that a pion carries the nuclear force is even less likely.

Decisive Experiment

When it was discovered in the 1970s that half of the proton's mass is not carried by quarks, scientists were astounded. When they recovered, however, they provided an explanation in hindsight whereby the missing mass was actually carried by gluons. Gluons, in their view, were the particles carrying the strong force, and they were trapped inside protons and neutrons, unable to exist in free form.

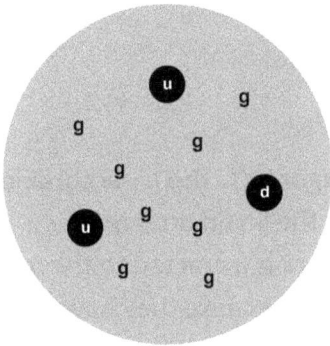

Figure 44. The proton according to QCD. Gluons carry the strong force as well as half of the proton's mass.

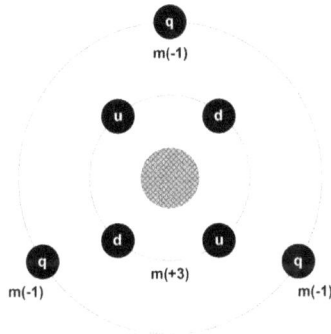

Figure 45. The proton according to the Liquid Drop Model (E model). Half of the proton's mass is situated in the core and in the internal quarks.

According to QCD, these gluons should also exist inside mesons, where they ought to keep the u quark together with the $\bar{d}$ antiquark inside π^+ pions. If QCD's explanation is correct, gluons ought to carry about one-half of the meson's mass inside π^+ particles as well.

According to the liquid drop model, however, the entire mass of the π^+ particle ought to be carried by quarks, because in this model gluons do not exist, and mesons do not have an inner core.

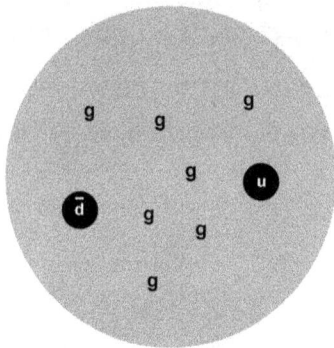

Figure 46. The π+ according to QCD. The quark is attracted to the antiquark by gluons.

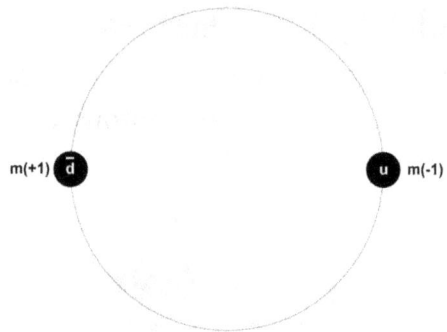

Figure 47. The π+ according to the liquid drop model. The quark is attracted to the antiquark because they have opposite strong charges.

At this point, therefore, all we have to do is conduct an experiment similar to the one completed in the 1970s, only this time using pions instead of protons.

Particle accelerators have been able to generate π^+ beams for quite some time. It's possible, for example, to have them collide with powerful electron beams and then perform an analysis similar to the one performed in the 1970s with protons and an electron beam.

If the results of the experiment show that the pion's mass is significantly larger than the total mass of the quarks that comprise it, it will therefore follow that the liquid drop model is incorrect.

And, if the results show that the pion's mass is similar to that of the quarks that comprise it, it will then follow that there are no gluons and that QCD is incorrect.

The experiment was suggested several years ago,[86] but it was never conducted.

---

86 E. Comay, A regular theory of magnetic monopoles and its implications, published in "Has the Last Word been Said on Classical Electrodynamics?" Rinton Press, NJ, 2004.

A List of Unexplained Phenomena So Far

At the end of each section I will present all problematic phenomena that have been examined, divided into two categories:

- Contradicting phenomena: Phenomena that are inconsistent with QCD predictions and that have not been explained to this day. Another class of contradicting phenomena includes phenomena that were known even before the advent of QCD, and for which there still exists no conventional explanation. Finally, another class of contradicting phenomena includes those phenomena that are explained by QCD or the Standard Model in a manner that contradicts the laws of physics.

- Suspicious phenomena: Other unexplained phenomena or phenomena that defy our expectations, and for which explanations were only given in hindsight.

Table 4. Problematic phenomena associated with inter-nucleon forces.

Contradicting Phenomena	Suspicious Phenomena
1. Nuclear attractive force	1. Half of the proton's mass not found in external quarks
2. Constant density inside the nucleus	

Contradicting Phenomena	Suspicious Phenomena
3. Similarity between nuclear and van der Waals forces (as indicated by the "potential graph")	
4. Why, in deuterons, protons and neutrons fail to combine into a single particle	
5. The EMC effect	
6. Nuclear tensor forces	
7. Strong CP problem	

By the end of this book we will have accumulated over twenty contradicting phenomena, and several other suspicious ones.

It's also interesting to examine if the alternative models are capable of explaining nuclear phenomena. It appears that both the D and E models are able to explain nuclear forces and the EMC effect. As for tensor forces, novel equations that describe strong forces will be required in order to see if experimentally measured tensor forces are compatible.

When it comes to the nuclear force operating between nucleons, it is common knowledge that QCD is facing a problem. QCD, however, is supposedly successful at explaining other phenomena. As we will see in the following chapters, QCD fares no better at explaining many other experimental results.

THE PHOTON AND THE STRONG FORCE

What Carries the Electric Force?

In the next series of chapters we will discuss the particle, the discovery of which was instrumental in establishing the foundations of quantum theory—namely, the photon. Before we can understand the role of this important particle in nature, and particularly its role in the function of strong forces operating inside nucleons, we will first examine the validity of several preconceptions.

One minor note before we delve into the details: the entire discussion presented in this chapter is not trying to show any flaws in QCD theory. It will, however, enable us in coming chapters to better understand the role of the photon in the strong forces that operate inside neutrons and protons.

The photon is defined by numerous textbooks and websites, like the Hebrew version of Wikipedia (2013), as "an elementary particle, the quantum of light, and all other forms of electromagnetic radiation, and the force carrier for the electromagnetic force...." The English version is slightly different and I will discuss it later.

This statement, that the photon is the "force carrier for the electromagnetic force," is actually quite new. Until the mid-twentieth century, the photon and the "electromagnetic field" were two clearly distinct concepts. The photon interacted with electric charges, but no one argued that it was the photon that "carries" the electric force.

We will begin by showing that one cannot simply accept the idea that photons are indeed the force carriers for the electromagnetic force.

Because of Its Spin

The positive charge inside atoms is located in protons that attract electrons in the atomic shell. If photons do carry the electromagnetic force, then the positive center would emit photons that are absorbed by electrons.

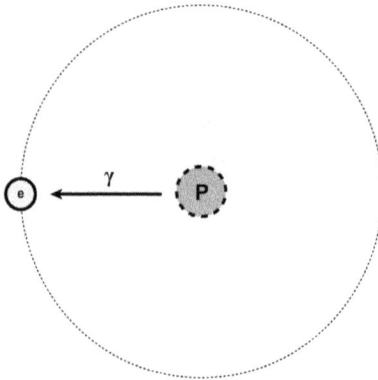

Figure 48. The proton and electron inside the hydrogen atom. Were the proton to emit a photon in order to hold the electron, and were the electron to absorb it, the electron's parity would change concurrently with that of the proton inside the nucleus. This, however, fails to occur.

However, photons are odd, spin 1 particles. When an electron absorbs a photon, it needs to comply with the law of conservation of angular momentum, as the entire atom reverses its parity (this topic is discussed in earlier chapters on spin). When an even atom with even parity electrons absorbs a photon, its parity becomes odd and vice versa. It therefore follows that if the positive center truly emits photons to the electrons in the atomic shell, it means that the parity of those electrons ought to change almost constantly. Instead, we know that parity is a stable property that is maintained by the atom so long as there is no external interaction. A hydrogen atom in its ground state, for example, has an even-parity electron. The hydrogen atom will remain in this state so long as it is not affected by external factors.

An examination of the nucleus reveals an even greater flaw. I will demonstrate it by analyzing a helium ion whose nucleus holds one electron. If the helium nucleus emits a real photon, then it must change its parity and its angular momentum as well. Moreover, since the ordinary helium nucleus has a zero angular momentum, it must shift to a state where it has an angular momentum of 1. Each of these changes forces the nucleus to shift into an energetic state much higher than its ground state, as nuclear energy is several orders of magnitude greater than that of electrons inside the atom. This effect rules out the possibility that real photons are involved in the operation of the force that keeps electrons bound to the nucleus.

Therefore, atomic nuclei cannot attract electrons by emitting real photons.

Virtual Photons

There are other contradictions to the simplistic acceptance of the idea that photons carry electric force that do not necessitate in-depth understanding. For example, let's examine a free electron at rest. The electron has a charge, and it exerts force on any other charge in the universe. How does it do this? When at rest, it cannot send photons in any direction because of the law of conservation of energy. Furthermore, because a free electron does not disappear within a short amount of time, and because it conserves its energy so long as it does not enter an electromagnetic field, it is unclear how it might be able to emit photons capable of wandering freely until they find an electric charge to harass. Some scientists attempt to solve this much more obvious problem by positing a "virtual" photon that is distinguishable from a "real" one.

The English Wikipedia "Photon" article is slightly different from the Hebrew article, as it addresses the concept of "virtual" photons.[87] The thing is, virtual photons are not real photons, and the question of whether virtual particles exist is more philosophical than scientific.[88]

The Feynman Diagrams

Whence, then, comes the idea that photons carry electromagnetic force?

In the 1940s and 1950s something known as field theory was developed. A significant part of that theory was called quantum electrodynamics, or QED. Field theory provided computational tools that afforded unprecedented accuracy when it came to electric forces.

---

87   en.wikipedia.org/wiki/Photon. "*A photon is an elementary particle, the quantum of light and all other forms of electromagnetic radiation, and the force carrier for the electromagnetic force, even when static via virtual photons.*"

88   en.wikipedia.org/wiki/Virtual_photons. "*...their 'reality' or existence is a question of philosophy rather than science.*"

The principal computational tool provided by field theory is called the Feynman diagrams, named after Richard Feynman, who developed them. Some Feynman diagrams include a photon that carries the electric force.

Do the Feynman diagrams imply that the photon indeed carries the electric force? Not exactly.[89] But Feynman used the photon equations to develop his diagrams. At some point as he developed his equations Feynman arrived at unphysical, infinite quantities. He "removed" these infinite quantities, without mathematical justification, in order to arrive at finite quantities. The finite quantities obtained were spectacularly commensurate with experimental findings.

The removal of infinite quantities is known as "renormalization." Dirac criticized the concept of renormalization by sarcastically arguing that we're allowed to remove miniscule quantities, but we shouldn't remove infinite quantities just because we find them uncomfortable.[90] Feynman himself noted in his book that he finds the whole process of renormalization uncomfortable particularly because it succeeded, because it is not mathematically justified, and because the scientific community prefers to be content with

---

89 profmattstrassler.com/articles-and-posts/particle-physics-basics/virtual-particles-what-are-they/. *"The Feynman diagram is actually a calculational tool, not a picture of the physical phenomenon."*

90 Helge Kragh, *Dirac: A scientific biography*, Cambridge University Press 1990, p. 184. *"Most physicists are very satisfied with the situation. They say: 'Quantum electrodynamics is a good theory and we do not have to worry about it anymore.' I must say that I am very dissatisfied with the situation, because this so-called 'good theory' does involve neglecting infinities which appear in its equations, neglecting them in an arbitrary way. This is just not sensible mathematics. Sensible mathematics involves neglecting a quantity when it is small—not neglecting it just because it is infinitely great and you do not want it!"*

this unmathematical process rather than look for a suitably consistent theory.[91]

The transformation of field theory into a consistent theory is perhaps the most important problem in the field of particle physics, despite, as Dirac and Feynman feared, the fact that every scientist is happy with its current formulation and prefers to disregard this inconsistency. What we can say in the meantime is that as long as there isn't a consistent theory whereby photons carry electric force, arguing that they do would be unjustified, especially when we find serious contradictions to this idea, as mentioned in the beginning of this chapter.

Schools

Today we can see what scientists think about the subject, since the rather irksome question of how photons are able to carry electric force is occasionally presented. It is not easy to find two scientists who give the same answer to that question. One can try dividing them into several groups: the first group is composed of physicists who actually believe that a real photon literally carries electromagnetic force; another group includes physicists who endorse the concept of virtual photons; the third group consists of those who

---

91   Richard P. Feynman, *QED, The Strange Theory of Light and Matter*, Penguin 1985, p. 128. "*But no matter how clever the word [renormalization], it is still what I would call a dippy process! Having to resort to such hocus-pocus has prevented us from proving that the theory of quantum electrodynamics is mathematically self-consistent. It's surprising that the theory still hasn't been proved self-consistent one way or the other by now; I suspect that renormalization is not mathematically legitimate.*"

do not clearly differentiate between virtual and real particles;[92] and the fourth group considers the wording of "virtual" photon an unfortunate mistake, as the carrying of electromagnetic force shouldn't be ascribed to particles at all.[93]

Let's end the discussion here, and agree that a real photon is not a particle that carries electromagnetic force.

Consistency: Who Needs It Anyway?

Since we've touched upon the topic of renormalization and the importance of a consistent theory, let's see where physics gets us when it abandons the need for consistent theories. We will see how the acceptance of renormalization as a legitimate mathematical tool might lead distinguished physicists to the realm of absurd.

In a short YouTube video[94] uploaded in January 2014 two physicists, Toni Padilla and Ed Copeland of Nottingham University, who specialize in string theory, can be seen. String theory is a branch of particle physics that involves thousands of scientists from all over the world. Opinions vary as to whether there are any physical

---

92   Scientific American, October 9, 2006, Ask the Experts. *"Are virtual particles really constantly popping in and out of existence? Or are they merely a mathematical bookkeeping device for quantum mechanics? Gordon Kane, director of the Michigan Center for Theoretical Physics at the University of Michigan at Ann Arbor, provides this answer. Virtual particles are indeed real particles..."*

93   profmattstrassler.com/articles-and-posts/particle-physics-basics/virtual-particles-what-are-they/. *"Perhaps unfortunately, this type of disturbance, whose details can vary widely, was given the name 'virtual particle' for historical reasons, which makes it sound both more mysterious, and more particle-like, than is necessary..."*

94   www.youtube.com/watch?v=w-I6XTVZXww.

insights that owe themselves to string theory. Two books[95],[96] published almost a decade ago discredited the theory as illegitimate.
One of them is amusingly called *Not Even Wrong*.

The video begins by claiming that it is about to prove that the sum
of the series:

$$1 + 2 + 3 + 4 + ...$$

is equal to -1/12. Does that sound ridiculous? Not according to
the two physicists in the video. One of them, Ed Copeland, even
mentions that contemporary physics uses this result for many
purposes, and Tony Padilla opens a book called *String Theory*,[97]
which presents the following equation as a legitimate expansion
of renormalization.

$$A = \frac{D-2}{2} \sum_{n=1}^{\infty} n , \qquad (1.3.31)$$

the factor of $D - 2$ coming from the sum over transverse directions. The
zero-point sum diverges. It can be evaluated by regulating the theory and
then being careful to preserve Lorentz invariance in the renormalization.
This leads to the odd result

$$\sum_{n=1}^{\infty} n \rightarrow -\frac{1}{12} . \qquad (1.3.32)$$

Figure 49. An excerpt from a book on string theory
that expands the concept of renormalization.

---

95   Lee Smolin, *The Trouble With Physics: The Rise of String Theory,
The Fall of a Science, and What Comes Next*, 2007.

96   Peter Woit, *Not Even Wrong: The Failure of String Theory and the
Search for Unity in Physical Law*, 2007.

97   Joseph Polchinski, *String Theory*, Cambridge Monographs on
Mathematical Physics, 2005. p. 22.

At that point, we are presented with "proof." Let's consider the following three sums:

$$S_1 = 1 - 1 + 1 - 1 + \ldots$$
$$S_2 = 1 - 2 + 3 - 4 + \ldots$$
$$S = 1 + 2 + 3 + 4 + \ldots$$

Padilla begins by "proving" that the first sum (S_1) equals ½. Why ½? Because sometimes the sum is equal to zero, sometimes to 1, and, when stretched to infinity, the sum is equal to the average of the two—namely, ½. "There are other ways to prove that this sum is a half, by the way," Padilla adds.

Anyone who attended a single semester of a mathematics course would know that the sum of this series, as it is defined in conventional mathematics, is *not* ½. In fact, its value cannot be defined. The definition of the sum of series was formulated by the French mathematician Augustin Cuachy almost two hundred years ago, but let's continue and see what these physicists are saying.

We will now see what the sum of $S_2 + S_2$ is, but we will add them together with a slight shift:

$$S_2 + S_2 = 1 - 2 + 3 - 4 + 5 - 6 \ldots$$
$$+ 1 - 2 + 3 - 4 + 5 \ldots$$

And, if we add each number in the top row with its counterpart in the bottom row, we will arrive at:

$$S_2 + S_2 = 1 - 1 + 1 - 1 + 1 - 1 \ldots$$

Which is equal to S_1, and so the value of $S_2 + S_2$ is ½, and so the value of S_2 is ¼. According to conventional mathematics, one cannot add the series terms when neither of them has a definite sum. But let's move on.

Here our "proof" continues. Padilla now calculates the value of $S - S_2$:

$$
\begin{array}{rrrrrrrrrrrrr}
S & - & S_2 & = & 1 & + & 2 & + & 3 & + & 4 & + & 5 & + & 6 & \ldots \\
 & & & - & 1 & + & 2 & - & 3 & + & 4 & - & 5 & + & 6 & \ldots \\
 & & & = & & & 4 & & & + & & 8 & & + & 12 & \ldots \\
 & & & = & 4 & \times & S & & & & & & & & &
\end{array}
$$

And from that the physicists conclude that because the value of S_2 is ¼, it follows then that the value of S is -1/12.

If this video was simply designed to offer some mathematical puzzle in which viewers are asked to find the mathematical flaw inherent in it, that's great. However, the scientists who presented it truly believe the mathematics they employ, and they mention how important these results are to physics and to understanding "twenty-six-dimension" physical theories.

We can develop a mathematical theory with laws that differ from conventional mathematics, but each such theory must be consistent—namely, it must be devoid of any internal contradictions. Let's now see if Padilla and Copeland's arguments contain any such internal contradictions.

The principal tool used for this proof is the shifting and addition of series. This is something we're allowed to do only when the sum of the series is defined (or when it converges, in mathematical terminology). Let's assume that this tool is valid and the value of S is indeed defined. We will now calculate the sum of $S + S - S - S$ and see what we get after a small shift. We should expect to get the same result—that is, zero. But let's see what really happens:

$$
\begin{array}{rrrrrrrrrrrrrr}
S & + & S & - & S & - & S & = & 1 & + & 2 & + & 3 & + & 4 & + & 5 & + & 6 & \ldots \\
 & & & & & & & & & + & 1 & + & 2 & + & 3 & + & 4 & \ldots \\
 & & & & & & & & - & 1 & - & 2 & - & 3 & - & 4 & - & 5 & \ldots \\
 & & & & & & & & - & 1 & - & 2 & - & 3 & - & 4 & - & 5 & \ldots \\
 & & & & & = & 1 & & & & & & & & & & & &
\end{array}
$$

What this means, according to our calculations, is that the sum of S + S – S – S is 1. If one adds all the series together without shifts one gets zero. The end result of this is that 1 is equal to zero.

This means—under the mathematical laws on which the video is founded, and the same could perhaps be said about parts of string theory—that one arrives at an internal contradiction; namely, that 0=1 and until string theory was invented no one had ever noticed it. By the way, one can shift series in other ways and arrive at other sums. I invite readers to amuse themselves, and prove that they can actually shift the series to arrive at any integer by adding these four series together.

Even though the occupation with this video seems rather trivial, it can still teach us a few things. Here's what I learned:

- When someone says that something is simple and self-evident, and that there are "many ways to show it," that would be a good time to start getting suspicious, even when this is said by someone who is capable of understanding worlds with twenty-six dimensions.

- Amazon offers science books that contain absurd ideas.

But the main lesson to learn here is that thousands of physicists have strayed so far off the path of conventional mathematics that they have no problem with formulating entirely inconsistent theories. In the case of renormalization, one employs it whenever an inconsistent theory leads to interesting results, and, in the case of string theory, inconsistency is excusable even if the theory is utterly useless.

The Photo-Strong Effect

Let's return to the basics for a moment and revisit the photoelectric effect. When a photon hits an electron, it interacts with it only if it is able to shift it from one energy level to another. Until the 1950s, it was a commonly held belief that photons only interacted with electric charges.

More than fifty years ago, however, scientists were amazed to discover that a high-energy photon beam interacts very strongly with protons and neutrons. According to quantum theory, the intensity of this interaction between the photon and the electric charge depended on the size of the electric charge. What scientists discovered, however, was that the interaction between photons and nucleons was a lot stronger than what one might expect from an interaction between a photon and the electric charge inside nucleons. Even more curious was the fact that the interaction between the photon and proton was very similar to its interaction with the neutron, even though the proton's electric charge is completely different from that of the neutron!

These phenomena astounded scientists and have remained unexplained to this day. Paradoxically, however, only a very few physicists are aware of their existence. This bizarre tale will also shed light on the evolution of QCD in the 1970s.

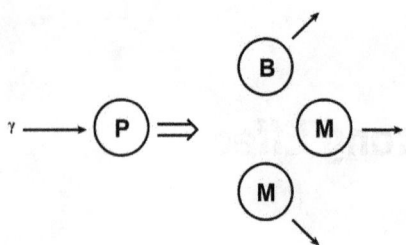

Figure 50. Inelastic interaction between an energetic photon and a proton. The interaction is much stronger than what one might expect from the electromagnetic interaction between photons and the electric charges inside the proton.

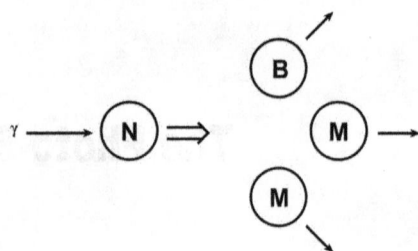

Figure 51. Inelastic interaction between an energetic photon and a neutron. This strong interaction is very similar to that of the photon and proton, despite the difference in electric charge between the proton and neutron.

Vector Meson Dominance (VMD)

A short time after this surprising discovery, physicist Jun John Sakurai came up with a revolutionary theory that could supposedly explain this phenomenon.[98] According to Sakurai, the photon was not a "pure" electromagnetic photon as quantum theory would have it, but rather a superposition of a pure photon and spin 1 mesons that are of odd parity (the photon is also an odd spin 1 particle).

This theory elegantly answered the question of why the interaction was so strong: some of the mesons that were scattered in eve-

---

98   J. J. Sakurai, *Theory of strong interactions*, Ann. Phys., **11** (1960).

ry direction as a result of the collision of the photon and nucleon belonged to the photon itself! Not only that, but since a large part of the interaction was ascribed to mesons that belonged to the photon, this explained why the interaction with the neutron was similar to its interaction with the proton.

Sakurai, who named his theory vector meson dominance, provided equations that predicted the value of the interaction between photons and nucleons.

If readers would like to know more about the subject, such properties of light are known in academic literature as the "hadronic properties of the photon."

VMD Discredited

Today we know that mesons consist of quarks and antiquarks. Quarks and antiquarks are massive particles; namely, they move at slower-than-light speed. It follows, then, that according to VMD photons are a mixture of a particle that always moves at light speed and particles that always move at slower-than-light speed.

Does that make sense? Not at all. According to special relativity and according to an article by Wigner,[99] a particle must belong to one of two groups: those that always travel at light speed and those that always travel at slower-than-light speed. This work by Wigner is considered to be incredibly important and profound, and is described as follows: "It is difficult to overestimate the im-

---

99 E. P. Wigner, *On Unitary Representations of the Inhomogeneous, Lorentz Group*, Annals of Math, **40**, 149 (1939).

portance of this paper, which will certainly stand as one of the great intellectual achievements of our century."[100]

Sakurai developed his theory before the advent of the quarks model, and at the time no one knew that mesons were made up of massive sub-particles.

Despite this fundamental "bug" in Sakurai's theory, it took thirty years to finally discredit it and remove it from textbooks. In the early 1970s it turned out that VMD's equations failed if one employed energies higher than those used in the 1960s. Sakurai improved his theory and eventually developed the concept of generalized vector meson dominance (GVMD). This new and improved theory provided novel equations that agreed with the findings known to him at the time.[101] In the 1980s, however, even stronger interactions were achieved, and GVMD's equations were once again found wanting. Sakurai died in 1982, leaving no one after him to improve his theory. Jerome Isaac Friedman put the final nail in the theory's coffin during his Nobel lecture.[102]

Curiously, not only was VMD removed from textbooks, but even the phenomenon itself ceased to be mentioned. Four contempo-

---

100 S. Sternberg, *Group theory and physics*. Cambridge University Press, 1994. p. 149.

101 J.J. Sakurai and D. Schildknecht, *Generalized vector dominance and inelastic electron nucleon scattering—the small ω region*, Phys. Lett., **40B** (1972) 121.

102 J.I. Friedman, *Nobel Lecture, Deep Inelastic Scattering, Comparisons with the Quark Model*, Rev. Mod. Phys. vol **63** 3 (1991). "*...this eliminated the model [VMD] as a possible description of deep inelastic scattering... calculations of the generalized vector-dominance failed in general to describe the data over the full kinematic range...*"

rary textbooks that should examine this basic phenomenon fail to address it even once.[103,104,105,106]

The Significance of the Electro-Strong Effect

If we ignore VMD theory and choose to seek another logical explanation to the photon's hadronic properties, we immediately notice the similarities to the photoelectric effect. In this case, let's consider the photoelectric effect of very low-energy photons that travel near atoms and fail to create an interaction with them if they can't shift the electrons to a higher level of energy. Similarly, the hadronic properties of the photon do not manifest themselves when the photon is at a relatively low energy. These properties only appear in very strong gamma rays.

A credible explanation would be that photons do not only create an interaction with electric charges, but also with strong charges (carried by quarks), which are much stronger than electric charges. The reason only very energetic photons create very strong interactions is that a substantial amount of energy is needed to shift protons and neutrons to a high enough energy level to produce and emit mesons. This explanation does not conform to QCD, which states that gluons and not photons are the particles responsible for the interaction with strong charges.

103 D.H. Perkins, *Introduction to high energy physics*, (4th ed. Cambridge University Press, 2000).

104 D.J. Griffiths, *Introduction to elementary particles*, (2nd, rev. ed. Weinheim: Wiley-VCH, 2008).

105 F. Halzen and A.D. Martin *Quarks and leptons*, (New York : Wiley, 1984).

106 Fayyazuddin and Riazuddin, *A modern introduction to particle physics*, (2nd ed. Singapore: World Scientific, 2000).

The hadronic properties of the photon are unexplained by the Standard Model.[107]

So what can this photo-strong effect indicate to us? It's certainly possible that the photon is a particle whose role in electromagnetic theory is similar to its role in a valid theory of strong interactions.

Figure 52. According to the photoelectric effect, when a photon is unable to shift an atom to a higher energy l evel it fails to create an interaction with it.

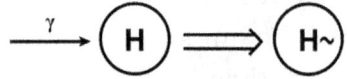

Figure 53. When a photon is able to shift an atom to a higher energy level, it is absorbed by the atom, which in turn becomes "excited."

Figure 54. When a photon is unable to shift a proton to a higher energy level, it fails to create an interaction with it.

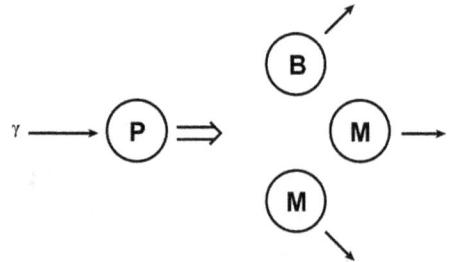

Figure 55. A photon is able to excite a baryon if it is sufficiently energetic.

107   H.B. O'Connell, B.C. Pearce, A.W. Thomas and A.G. Williams, *Rho-omega mixing, vector meson dominance and the pion form-factor,* Prog. Nucl. Part. Phys. **39** (1997) 201–252. "*No direct translation between the Standard Model and VMD has yet been made.*"

A strange anomaly we find here is that the photoelectric effect is a basic phenomenon known by every physics student, whereas its strong force counterpart is known only to a select few students and active physicists.

The Three Jet Event

In 1979 (or 1978, as we will clarify soon), an experiment was conducted that resulted in the discovery of the three jet event, considered to be irrefutable evidence for the existence of gluons, and therefore to the veracity of QCD. Here is how the results of the experiment are summarized by John Ellis:

> ...the gluon finally joined the Pantheon of established particles as the first elementary boson to be discovered after the photon.[108]

I slightly paraphrased Ellis's summary to avoid using new terminology.

Let's examine the connection between this exciting announcement and the actual results of the experiment. First, however, I should note that gluons are unable to exist in free form, and therefore evidence for their existence is strictly circumstantial. This is why this experiment is so important for transforming the gluon from a theoretical concept to a particle that is accepted by physicists.

Bremsstrahlung

The energetic collision of two electric charges allows them to get very close to each other. Under such circumstances a photon may

---

108 cerncourier.com/cws/article/cern/39747.

be emitted, producing a phenomenon known as bremsstrahlung. This is a known phenomenon in the field of electromagnetic forces. Quantum theoretical calculations show that the probability of a bremsstrahlung event is very small when it comes to electric forces, and it is proportional to the strength of the charge to the sixth power.

Bremsstrahlung and Strong Forces

QCD scientists have assumed that the same phenomenon applies to strong forces—namely, as quarks get closer to one another, a photon-analogous particle may be emitted as a result.

As far as QCD is concerned, the photon-analogue in this case would be a gluon. The gluon is a spin 1 particle, even though many of its other properties are not analogous to the photon's—the gluon carries a strong charge while the photon does *not* carry an electric charge; the gluon carries the strong force and the photon does *not* carry the electromagnetic force, as we've already seen, and the photon can be a free particle but the gluon cannot exist as a free particle because it is not "white," according to QCD.

In experiments conducted in 1978 using the PLUTO detector and then later in 1979 using the PETRA particle accelerator, the phenomenon corresponding to the electromagnetic bremsstrahlung was indeed discovered.

PLUTO and PETRA were used to emit an energetic electron beam and have it collide with an energetic positron beam. The electron collided with the positron at tremendous energy, and they annihilated each other as one would expect from the collision of a particle and antiparticle, and the vast amount of energy generated at their collision point resulted in the formation of a quark pair.

The strong bremsstrahlung phenomenon occurred at the moment of that quark pair's formation. It turns out, as was predicted by QCD scientists, that apart from the two quarks, which emitted jets that flew in two directions, another particle had formed, which flew in a third direction and created an interaction with nearby hadrons. The two quarks and this other particle were the source of those three jets.

According to QCD proponents, the fact that the other particle interacted with nearby quarks proved that it was a gluon. QCD scientists were not satisfied with just that, however, and work published by Inga Karliner and John Ellis[109] enabled scientists to eventually measure the third jet's spin. Scientists were able to demonstrate that the third jet was indeed formed by a spin 1 particle, as should be expected from a gluon.

The success of this experiment and the ability to calculate the spin of the particle that created the third jet should be considered a resounding triumph. This I write, for a change, without the sarcasm that characterizes other parts in this book so far.

The Third Jet

If readers read previous chapters, they probably realize what I'm getting at by now, right? What does the three jet event actually tell us? That the strong and electric forces share yet another common trait—namely, bremsstrahlung. And now, all we need to do is find out the properties of the particle that created the third jet. Our two candidates are the gluon and the photon. As one can't

---

109 J. Ellis and I. Karliner, *Measuring the spin of the gluon in e+e– annihilation*, Nuclear Physics B, vol **148**, issues 1-2, (1979) p. 141–147.

simply ask the particle that created the third jet whether it's called a gluon or a photon, let's try and see which of the two is a more natural candidate:

- The photon interacts with electric and strong charges.

- The photon does not possess an electric or strong charge.

- The photon does not carry electric or strong forces.

These facts suggest that the photon affects electromagnetic forces in the same way it affects strong forces. Considering the fact that these properties can induce the bremsstrahlung effect in an electromagnetic field, they should also be able to induce that effect when it comes to strong forces. Therefore, the photon is the perfect candidate for being the particle that carries the third jet. By the way, both the photon and the gluon are spin 1 particles, so this particular property does not result in a distinction between the two.

And what about the gluon? Well, let's see:

- The photon interacts with electric charges and the gluon interacts with strong charges.

- The photon does not possess an electric charge, but the gluon possesses a strong charge.

- The photon does not carry electric force, but the gluon does carry strong force.

- The photon is a free particle but the gluon cannot exist as a free particle.

If we look for a perfect analogy between electromagnetic bremsstrahlung and strong bremsstrahlung, we find that the photon is a much more natural candidate for being the carrier of the third jet. All the photon's electric properties as they pertain to electric force also apply to strong force. The strong-force gluon, however, is not analogous to the electromagnetic-force photon when it comes to most of the properties we reviewed, and therefore it is a less natural candidate for being the third jet's carrier.

On Science and Drama

The perfect physicist is a dry, skeptic, pathetically pedantic character. The perfect physicist would never portray electrons as billiard balls. He or she would prefer to write down the Dirac equation, and would consider it to be much simpler to understand. He or she would never say that a certain theory is "true." At most, he or she would be content with saying that, to the best of his or her knowledge, the theory is yet to be refuted. In short, the perfect physicist is not a person anyone would want to be stuck with on a desert island, where the risk of dying of boredom would far exceed any other danger.

John Ellis is a rather colorful physicist who, even at his advanced age, still drives a motorcycle to work. Immediately after the third jet was discovered in 1979 at the PETRA accelerator, he dramatically announced the discovery of the gluon during the annual CERN convention.

Another group of scientists, which had made the same discovery in PLUTO a year earlier, was content with writing an article that stated that the third jet was commensurate with QCD predictions, and that was that.

If one had to choose a person to be stuck on a desert island with, the choice is rather obvious. What isn't obvious, however, is who was actually right in this case.

A Summary of
Unexplained Phenomena Thus Far

We will now present a list of all the problematic phenomena found in this book's chapters on photons.

Table 5. Problematic phenomena associated with the interaction between photons and quarks.

Contradicting Phenomena	Suspicious Phenomena
8. Photons create an interaction with nucleons that is far stronger than what one might explain with an electric charge.	
9. The interaction of photons with protons is very similar to their interaction with neutrons.	

DYING OUT OR GETTING STRONGER?

Asymptotic Freedom

QCD's equations were published in 1972 by Gell-Mann and Frisch, and they describe the forces that operate between quarks. An analysis of these equations showed that, if QCD's equations indeed describe the strong force that operates between quarks, then the strong force possesses an uncanny property that is unlike any other force in nature: the greater the distance between quarks, the stronger the force that operates between them. This property is known as "asymptotic freedom," and it was discovered in 1973 by David Gross, Frank Wilczek, and David Politzer.

The discoverers of this property were awarded a Nobel Prize for their work, and it is of great significance for the subject matter of this book, as the discovery actually indicates that *if the strong force that operates between quarks becomes stronger at shorter distances, then QCD is false.*

And, therefore, in order to refute QCD, it's enough to show that the force operating between quarks becomes greater as the distance between them grows smaller, in a manner contrary to asymptotic freedom and similar to any other force in nature.

Before we observe three and half different ways in which the strong force defies asymptotic freedom, we will first tell why scientists fell in love with asymptotic freedom and were so keen to accept it as gospel, to the point that some view it as proof of QCD's veracity.

Quark Confinement

To this day—despite countless collisions between protons and other energetic particles, which have managed to produce a collision between a single quark and a proton—not a single quark has ever been removed from a proton. This phenomenon is known as "quark confinement."

According to QCD, the quark's confinement inside the proton is a natural outcome of asymptotic freedom: the greater the distance between a quark and other quarks, the stronger the force that keeps it in place and prevents it from fleeing the proton. Instead, the colliding quark releases energy by creating a quark-antiquark pair, and this pair escapes the proton and allows the collided quark to restore its original energy level.

We should note, however, that quark confinement is possible even in the absence of asymptotic freedom: it can also be achieved by a "strong enough" force that keeps them in place. Meson formation, as it were, is actually quite easy to accomplish; it only requires investing less than 140 MeV. If a quark's removal requires thousands of MeV, then quark-antiquark pairs will always form before we are able to remove the quark from the proton. This explanation was proposed as early as the 1960s.[110]

Later in this book we will see why QCD's explanation of confinement is problematic, since experiments show that it does not apply to mesons—namely, a quark-antiquark pair *can* escape the proton quite easily.

---

110   P. N. Bogolioubov, Ann. Inst. Henri Poincaré, 8, 163 (1968).

The Landau Pole

In the 1950s, Lev Landau, Alexei Abrikosov, and Isaak Khalatnikov[111] discovered that, according to field theory equations, the force operating between electric charges became infinite when they were situated at a very short distance from one another. This posed a theoretical problem, meaning that field theory equations, which represent the forces operating between electric charges, do not apply whenever the charges are exceedingly close to one another.

This theoretical problem stems from the fact that the force between electric charges increases as the distance between them decreases. The problem does not apply to strong forces as described by QCD equations, since according to QCD asymptotic freedom enables the force between quarks to diminish as the distance between them decreases.

Some scientists argued that the fact that this problem does not exist in QCD is yet more proof that QCD accurately describes strong forces.

What are these scientists actually saying? Everyone agrees that the electric force diminishes with an increase in distance. Everyone also agrees that field theory equations that describe phenomena related to electric force are problematic—that is to say, they must be revised. But, instead of revising the equations, scientists argue that strong forces ought to behave differently than what is known about electromagnetic forces, simply because we have thus far failed to produce consistent equations for one aspect of the electromagnetic theory.

---

111   L. D. Landau, A. A. Abrikosov, and I. M. Khalatnikov, *On the elimination of infinity in quantum electrodynamics*, Dokl. Akad. Nauk SSSR 95, 497, 773, 1177 (1954).

That doesn't sound like a very strong argument, now, does it? But in an interview I had with a scientist who supports QCD and who has published hundreds of articles on the subject, I presented him with the following question: "Why do you believe that QCD must be true?" For several hours he presented me with some ten different arguments, one of which was the Landau pole. Two other arguments involved the three jets event and the fact that there exist particles that contain three quarks whose spins are all pointing in the same direction. It's interesting to note that the most salient weak points of QCD were used by that scientist as evidence proving the theory he advocates.

And one last note about the Landau pole: more than a decade ago I was sued by the software company that had employed me several years earlier. The indictment included a series of unfair acts I had allegedly committed, all of which were completely unfounded. The arguments found in the indictment were false and easily refutable, but the effect brought about by this deluge of accusations seemed quite convincing to those who didn't bother with the details. Similar to the argument that the Landau pole provides proof of the veracity of QCD, the indictment also included several claims that were so lacking in internal logic that it was almost impossible to address them.

The main lesson I learned from this lawsuit is that a collection of poorly substantiated claims, and the use of inconsistent arguments, are unable to support a theory, and actually achieve the opposite by undermining its reliability.

Sheldon Glashow:
"The Thorn in the Side of QCD"

In 2007 Alan Krisch,[112] a known experimental physicist, had published a summary of experiments conducted by his team since the 1970s. In this chapter we will try to understand what these experiments were about, and how the scientific community addresses their findings.

Polarized Protons

Let's recall what we've learned about spin. When a beam of spin-bearing particles such as photons, protons, and electrons is emitted, the direction of the spin is usually random, and in fact the beam is normally a mixture of spin-bearing particles that face every direction permitted by the laws of physics.

A beam is referred to as "polarized" if the spin of each of its constituent particles is facing the same direction. Nature contains several examples of polarized light, but polarized proton beams are never found in nature, and they are in fact quite hard to pro-

---

112 Alan D. Krisch, *Hard collisions of spinning protons: Past, present and future*, The European Physical Journal **A 31**, 417-423 (2007).

duce even through the use of particle accelerators. The reason for that is that when one accelerates polarized protons with strong electromagnetic fields, one creates a "disturbance" that affects the precise directions found inside the polarized beam, thus causing it to lose its polarity.

Krisch's team at the Michigan Spin Physics Center had developed a specific technique that allowed it to emit polarized proton beams at unprecedented energy levels.

The Experiment: Collision between Polarized Protons

In an experiment conducted by Krisch's team in 1977 to 1978, a polarized proton target was bombarded with a strong beam of polarized protons. The experiment measured the level of interaction between the protons when the spin was pointing in the same direction, as opposed to the interaction between protons with opposite spins.

The results of the experiment astounded the scientists.[113],[114] The scientists thought that there would be no significant difference between the collision of protons with parallel spins and that of protons with opposite spins, but it turned out that the interaction between parallel-spin energetic protons was up to four times stronger than that of opposite-spin protons.

---

113   Ibid. *"people were totally astounded."*

114   *Proton spin surprise,* Science News, vol **112**, 1977. p. 196. *"That is, rather simply, that protons bounce well off each other when their spins are parallel...When the spins are anti-parallel...the protons don't even seem to notice each other. They appear to pass right through each other as if they were transparent. Bang! Wow! Balloon full of question marks."*

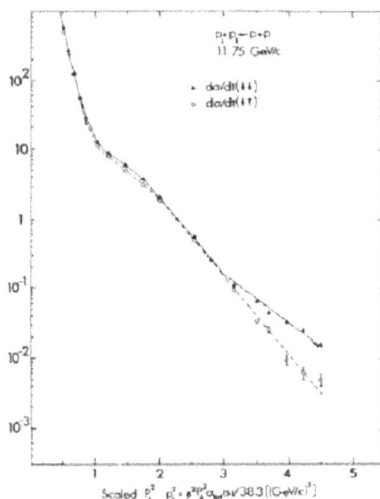

Figure 56. At high energy, the interaction between parallel-spin beams is four times stronger than that between opposite-spin beams (logarithmic scale).

Sheldon Glashow, who won the Nobel Prize in 1979 for his contribution to the Standard Model, called this curious finding "the thorn in the side of QCD."[115]

Krisch's group was sent to conduct several other experiments under different conditions in order to confirm these results. Every single experiment produced similar findings, all of which stand in contradiction to QCD.[116, 117] After the last experiment of this kind

115   Ibid (AD Krisch).

116   Cameron PR, Crabb DG, DeMuth GE, Khiari FZ, Krisch AD, Lin AM, Raymond RS, Roser T, Terwilliger KM, Brown KA, Danby GT, Ratner LG, O'Fallon JR, Peaslee DC, Roberts JB, Bhatia TS, Simonius M, *Measurement of the analyzing power for p+p*, Phys Rev D 32 (11) Part Fields. (1985) p. 3070–3072.

117   DG Crabb, I Gialas, AD Krisch, AMT Lin, DC Peaslee, RA Phelps, RS Raymond, T Roser, JA Stewart, KM Terwilliger, KA Brown, GT Danby, FZ Khiari, LG Ratner, JR O'Fallon, G Glass, *Measurement of Spin Effects in p↑+p↑ →p+p at 18.5 GeV/c*, Phys. Rev. Lett. 60, (1988) p. 2351–2354.

was conducted in the 1980s, Krisch found that it was impossible to obtain the financing needed for other experiments in this field.

These findings remain unexplained to this day. In a lecture given in 2005, Stan Brodsky referred to this phenomenon as "one of the unsolved mysteries in hadronic physics."[118]

Krisch sadly and humorously described the aftermath of his experiments by saying that experiments with polarized protons have become "unpopular in some circles."[119]

This important experimental finding, like many others that contradict the Standard Model, is also never mentioned in particle physics textbooks.

What Is This About?

I assume readers are not very likely to understand what it was that made Krisch's discovery so jaw-dropping to QCD proponents.

What's interesting to note here is that a similar finding was already obtained in the mid-twentieth century that relates to electron-proton collisions. In this case we are dealing with processes that take place between electric charges, and that stem from an equation known as the Rosenbluth formula. However, the Rosenbluth formula works well when applied to electromagnetic forces, as the force that operates between electric charges decreases as the distance between them grows.

The problem is that the forces operating between energetic protons are mainly strong forces, whereas according to QCD the

---

118 Ibid (AD Krisch).

119 Ibid (AD Krisch). *"One result of our experiments was to make them unpopular in some circles."*

strong force ought to diminish as the quarks comprising the protons draw closer to one another, and therefore the Rosenbluth formula does not apply to QCD.

Had the strong forced behaved in a manner that "parallels" the electric force, and had we seen "strong force pairs" that correspond to electric forces and magnetic dipoles, and had the strong force, similar to the electric force, grown weaker as the distance between the strong charges grows, then Krisch's finding would have been naturally accepted and explainable in a similar fashion to what we have known for over fifty years about electromagnetic forces.

Upon the publication of Krisch's team's findings, scientists resolved to confront the data. Krisch's team published a series of scientific articles signed by sixteen or seventeen team members.

In contrast, QCD proponents had an almost knee-jerk reaction. They classified the phenomenon as yet another one that would require us to wait patiently until an adequate explanation could be found.[120],[121]

It seems that where particle physics is concerned, infinite patience is a must-have commodity.

During the 2000s another group conducted additional experiments with polarized proton beams and this is their summary:

---

120 I. Peterson, *Proton spin plays key role in smash hits*, Science News, vol 138, November 3, 1990. *"Some of the claims, that these experiments violate QCD are gross overstatements," says Francis E. Close of University of Tennessee in Knoxville. "They show interesting phenomena, but this is the sort of dynamics that QCD theory isn't equipped to handle yet."*

121 Ibid. *"QCD has many successes, which you can't ignore," adds Charles Y. Prescott of the Stanford Linear Accelerator Center. "The theory's inability to explain the Brookhaven results is a problem for QCD but not necessarily a failure."*

The transversely polarized proton-proton data from RHIC have provided several surprises to the field, all of which remain unexplained.[122]

After these findings failed to initiate any real discussion on the validity of QCD, Krisch published a new summary and included a comment that every scientist is supposed to know:

There is a BASIC PRINCIPLE OF SCIENCE: If a theory disagrees with reproducible experimental data, then it must be modified.[123]

We can conclude that particle physicists are unaware to this basic principle of science.

---

122 Christine A. Aidala, *Spin in Hadron Reactions*, AIP Conf. Proc. 1149, 124 (2009)

123 A.D. Krisch, *Hard Collisions of Spinning Protons: History & Future*, Arxiv, (2010). The capitalization is in the original text.

The Attraction to the Center

The structure of the atom is successfully described by quantum mechanics in a way that is accepted by all physicists. Quantum mechanics provide equations that depict various properties of atoms with spectacular precision.

As for the internal structure of the proton, however, we must rely on measurements rather than theoretical equations.

Where Do We Normally Find Electrons inside the Atom?

Let's consider the simplest of all atoms, the hydrogen atom: it has one proton and one electron bound to it. Consider the ground state of this atom. Where are we more likely to find the electron, farther from the proton or closer to it?

According to quantum theory, the probability of finding the electron closer to the proton is much greater than the probability of finding it farther away from the proton. This property is caused by the electron's wave function, and it is explained by equations that describe the electric force that operates between the electron's charge and the proton's charge as a function of the distance between the two charges. Since the electric force is stronger when the electron is closer to the proton, the electron is more likely to be found near the center.

Quantum mechanical equations indicate that the probability of finding the electron as a function of its distance from the proton decreases exponentially (for readers unfamiliar with the term, see Figure 57).

Proton Form Factor

And what of the quarks inside the proton? Where do we normally find them? Are they more likely to be found near the proton's center or at its peripheral regions? In order to measure the distribution of quarks inside the proton, physicists measure a quantity known as "form factor." This factor allows us to calculate the distribution of quarks inside the proton.

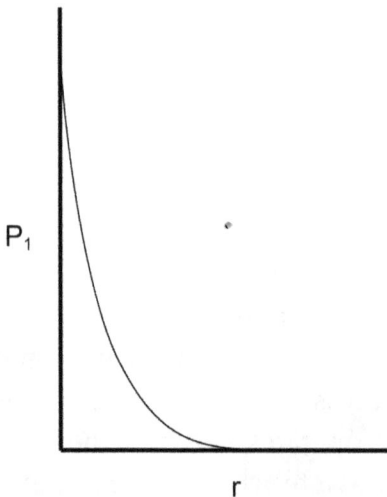

Figure 57. The distribution of the electron inside a hydrogen atom as a function of its distance from the center.

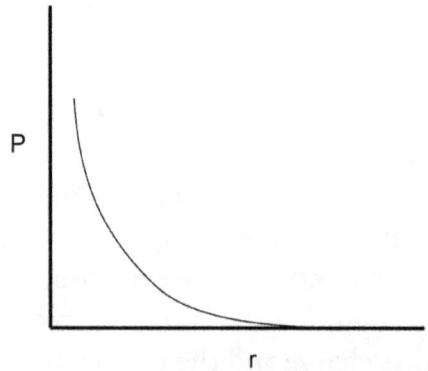

Figure 58. The distribution of quarks inside the proton as a function of their distance from the center.

Measurements results can be found in Figure 58. It turns out that, similar to electrons in the atom, quarks also tend to be found near the center, and here too we observe an exponential distribution! The obvious explanation for this is that the force operating between quarks decreases as a function of distance, just like the electric force operating on electrons inside an atom.

The data seen here on protons have been known to science since the 1970s. In a textbook authored by Perkins in the 1980s[124] readers will find a comprehensive calculation of quark distribution, but, for some reason, it omits the obvious conclusion, which is that the strong force operating on quarks decays in the same way as the electric force does.

Perhaps this conclusion is never mentioned because that would contradict QCD and the concept of asymptotic freedom, according to which the strong force supposedly decays in an opposite fashion to the electric force. In this regard Perkins is no different from other physicists who write technical articles that concern this scientific discipline: an examination of nuclear and particle physics textbooks would reveal that they fail to include a single direct confrontation between the author and QCD.

---

124 D. H. Perkins, *Introduction to High Energy Physics* (Addison-Wesley, Menlo Park, CA, 1987). p. 194–196.

The Radius of Baryons

Here we will examine the connection between the radius of a bound particle and the mass of its constituent parts. Radii are not easily defined in quantum mechanics, as each particle is actually a wave, and so we have to make do with an average value that will provide a description of the volume that contains the bound particle. The quantity we normally use is known as "charge radius," and, if any math buffs out there are interested, it is defined as the square root of the integral of the charge density multiplied by the radius squared.

Muonic Hydrogen

The hydrogen atom is made up of a proton and an electron. It's interesting to see what would happen if we replaced the electron with a similar particle known as a muon. The hydrogen-like atom in this case is composed of a proton and a muon, and is referred to as "muonic hydrogen." The muon has the same electric charge as the electron, and it too is an elementary particle that conforms to the Dirac equation. However, the muon is much heavier than the electron. The muon is not a stable particle, and it decays within about one millionth of a second.

In particle physics, a millionth of a second is actually a very long period of time, and so it allows us to measure quite a few of the muon's properties (and, as we will see later, to produce muon beams). Using particle accelerators, it is possible to produce muonic hydrogen, and even measure its radius.

It turns out that the radius of muonic hydrogen is substantially smaller than that of ordinary hydrogen. This property is also derived from quantum mechanical equations, and it stems from the fact that the muon is more strongly bound to the proton the nearer it gets to it. Because the muon is much heavier than the electron, the proton must pull it with greater force, and it is therefore situated closer to the center.

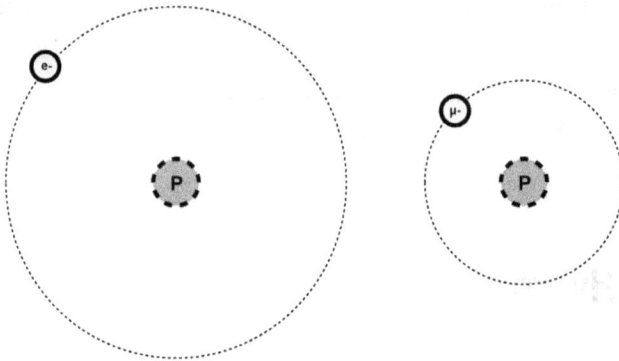

Figure 59. Muonic hydrogen (on the right) has a smaller radius than that of ordinary hydrogen (on the left).

In conclusion, the smaller radius of heavy particles bound by electric force stems from the fact that electric force decays as the distance between the charges grows.

So far everyone agrees. Now let's see how the strong force behaves.

Heavy Baryons

Protons and neutrons are the most common of baryons, and they both have similar mass. The proton has three *uud* quarks and the neutron has three *ddu* quarks. The *d* quark is similar in mass to the *u* quark, but other quarks, such as the *s* quark, are much heavier. A baryon of the sigma family is obtained when an *s* quark appears instead of one of the *u* or *d* quarks inside the proton (or neutron). These baryons decay very rapidly, and therefore it is not easy to measure their radii.

Nevertheless, a Σ^- particle's radius was measured in the late nineties,[125] and it appeared to be smaller than the radii of protons and neutrons.[126] A Σ^- particle is composed of the quarks *sdd*.

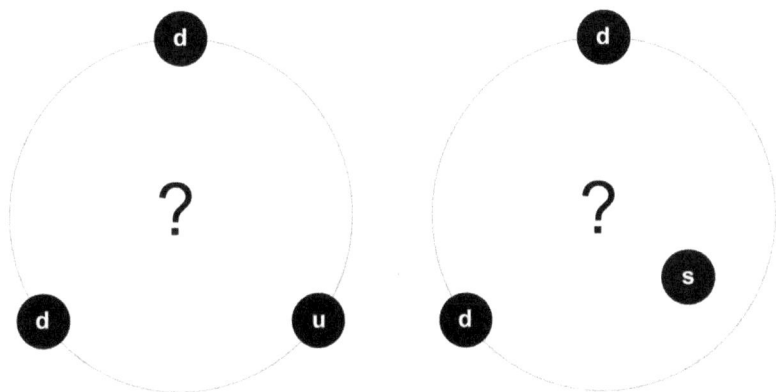

Figure 60. The radius of the $\Sigma-$ particle (on the right) is smaller than that of the proton or neutron.

The Σ^- particle is identical to the neutron in every respect apart from its having a heavier *s* quark that replaces the neutron's *u* quark. Therefore, were the strong force to decay with the increase in distance, as is the case for electric force, then the smaller radius

125   Ivo Eschrich, *Measurement of the $\Sigma-$ charge radius at SELEX*, arxiv. org, 1998.

126   K. Nakamura *et al.* (Particle Data Group), J. Phys. G 37, 075021 (2010).

of the Σ^- particle would have been explained in a manner that is analogous to the smaller radius of muonic hydrogen.

But the asymptotic freedom of QCD theory means that strong forces grow as the distance between the particles increases. Therefore, the relatively smaller radius of the Σ^- particle appears to contradict both asymptotic freedom and QCD.

An identical phenomenon was found in mesons. The radius of K^+, for example, is smaller than that of π^+. The π^+ meson is composed of the quarks u and $\bar{d}$, whereas K^+ is composed of u and $\bar{s}$—namely, the $\bar{d}$ quark is replaced by the heavier $\bar{s}$ quark, and, therefore, mesons also behave as one would expect from a force that decays with an increase in distance, as is the case for the electric force.

QCD proponents are not going to give up easily, and the fact that to this day all the radii of heavy-bound particles have turned out to be smaller than the radii of corresponding lighter-bound particles seems to provide them with no indication that they are dealing with a law of nature. As far as they're concerned, these results are completely incidental and in no way indicate a general rule.

For example, a few years ago a QCD-based prediction was published that said that the radius of the Σ^+ particle should be similar to that of the proton.[127] The radius of this particle is yet to be measured. I wonder what explanation will be given after the Σ^+ particle's radius is identified, as it joins the growing collection of findings that contradict the theory.

---

127   P. Wang, D. B. Leinweber, A. W. Thomas and R. D. Young, *Chiral Extrapolation of Octet Baryon Charge Radii*, Phys. Rev. D 79, 094001 (2009).

Cross Section

Cross section is a rather technical term, but it is also one of the most important scientific tools for studying the structure of nucleons. If we can understand how to read graphs that represent the cross section, we will be able to comprehend three different phenomena that contradict QCD. The more we progress in our analysis of the cross section graph, the more conspicuous the contradictions we find become.

For readers who would rather not delve into the details, please forgive the occasional use of mathematical terms in this brief chapter. In the next chapter we will consider other phenomena that require much less quantitative and much more qualitative explanations that describe the cross section graph.

Energy and Effective Distance between Colliding Particles

The cross section graph is derived from experiments that involve colliding particles. Let's remember that each particle is in fact a wave, whose wavelength shortens the higher the energy of the moving particle.

As a result, the higher the energy of the impinging partner, the shorter the effective distance of its interaction with the target particle. This can be visualized as follows: let's imagine the particle as a "padded" object. A higher energy means thinner "padding." Consequently, particles collide with each other from a greater distance the thicker the padding gets, and this happens when the colliding particles are less energetic.

The graph we will examine in this chapter, and in the chapters included in the next part of this book, describes the interaction between two proton beams as a function of their energy (speed). This graph is known as the cross section graph. In this graph, the more scattering events we observe, the larger the scattering cross section becomes.

The Cross Section Graph of Proton-Proton Collisions

One of the most important parameters in particle-particle collisions is the energy of the particles at the time of their collision. In order to add new points to the graph, we need to generate higher and higher proton beam energies. The graph referenced here[128] wasn't completed in a day, and it represents what was known in 2009. In the seventies, for example, scientists were unable to generate proton beams with the same energy that is possible today, and so forty years ago only the part on the left side of this graph was available.

---

128   C. Amsler *et al.* (Particle Data Group) Physics Letters B667, 1 (2008) see p. 12 in pdg.lbl.gov/2009/reviews/rpp2009-rev-cross-section-plots.pdf.

Incidentally, the points on the right end of the graph describe very energetic collisions. Unlike the rest of the points on the graph, these points have been obtained from measurements of cosmic rays. Measurements of cosmic rays are much less accurate, but you can see that they fit into the general trend shown in the figure.

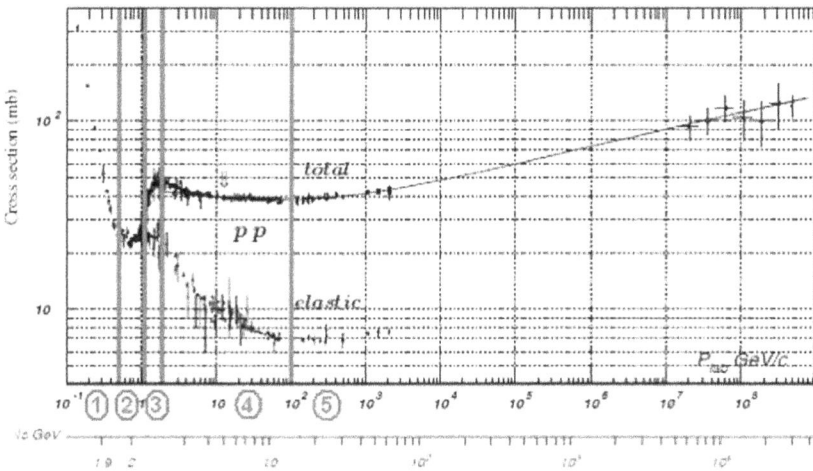

Figure 61. Energy dependence of the cross section of proton-proton collisions, divided into five different segments.

I divided the graph into five segments marked at the bottom of the graph to simplify our analysis of them. Segment 1, situated on the leftmost side of the graph, represents proton-proton collisions at low energy, and therefore the effective collision distance between the protons is relatively large. In this segment the interaction that occurs when the protons collide stems from the electric forces that operate between them. Notice that the graph's slope appears to be an almost straight line (the graph uses a logarithmic scale). The equations that describe electric force are the reason for this behavior.

Let's depict the process that occurs in this segment in a figurative and accessible fashion: on one hand, the interaction between protons diminishes as the amount of energy grows. This happens because the "thickness" of the proton is smaller when energy is higher, and therefore it "hits" the other proton with a lower probability (the word "hits" is somewhat misleading since the protons are actually waves). On the other hand, when protons collide with one another, the effective distance between them at the time of the collision is smaller, and their interaction is stronger because the electric force grows stronger as distance decreases.

What we see here then is a combination of two opposing factors: one leads to the graph's decay, and the other restrains that decay. In the case of a scattering process caused by electric forces, the cross section graph decays as collision energy increases, despite the fact that the electric force *intensifies* with the decrease in the inter-particle effective distance. According to quantum theory equations, the graph's gradient is determined by the fact that the electric potential is inversely proportional to the distance between two charges.

In Segment 2, the effective distance between the colliding protons is small enough that the nuclear force operating between them is able to manifest itself. This force is much stronger than the electric force, and its influence completely stops the graph's decay.

Segment 3 is characterized by higher energy and smaller effective distances between the colliding protons. In this segment we find that the graph is divided into two, a division we will discuss in later chapters. At this point let's only look at the bottom branch of the divided graph, which represents elastic collisions. Here we see another increase instead of further decay, which is caused by the strong force operating between the quarks of one proton and those of another.

Segment 4 is the one that should interest us the most in this chapter. For this energy the effective distance is so small that the process is a superposition of quark-quark scattering. Here we observe the force operating between the quarks of one proton and those of the other. We see that the graph falls relatively moderately when compared with Segment 1.

But, if asymptotic freedom exists, the force between quarks should *decay* with an increase in energy and *decrease* in the effective distance between the quarks of one proton and those of the other. Had the force between the quarks actually decreased, then the graph we see in Segment 4 would have decayed at a steep slope that should be stronger than the one we see in Segment 1. This means that the experimental results of proton-proton scattering actually refute the concept of asymptotic freedom espoused by QCD.

By the way, had there been only three quarks inside the proton without additional quarks or a core, and had the force between them behaved similarly to the electric force, then the graph we see in Segment 4 would have fallen at an identical angle to that of the gradient found in Segment 1. But, as we will see in the following chapters, there are other factors that further restrain this decay. In these chapters we will also examine Segment 5 of the graph, which contains the most salient contradiction to QCD.

A Summary of
Unexplained Phenomena Thus Far

We will now present a list of all the above mentioned phenomena that are at odds with asymptotic freedom.

Table 6. Problematic phenomena
that contradict asymptotic freedom.

Contradicting Phenomena	Suspicious Phenomena
10. High-energy collision of polarized protons with opposite spins display a weaker interaction than protons with parallel spins (Krisch's experiment).	2. All particles that contain an *s* quark instead of a *d* or *u* quark have a smaller radius.
11. The probability of finding a quark near the center of a nucleon is greater and is similar to the electron in a hydrogen atom.	
12. The cross section graph of elastic proton-proton collision decays rather slowly at energies where the quarks of one proton operate on the quarks of the other.	

SOMETHING MASSIVE INSIDE

Protons and Billiard Balls

When two particles collide without forming new particles as a result, the collision is referred to as elastic. When such a collision occurs, the particles behave in a manner akin to the collision of two balls on a pool table.

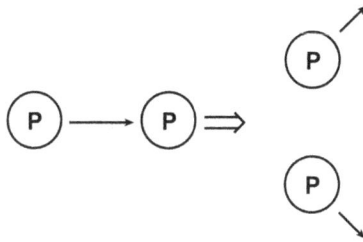

Figure 62. Elastic collision between two protons.
No new particles are formed.

When particles collide at higher energies, the most common outcome is inelastic collision, which causes the formation of new particles as a result of the collision's energy.

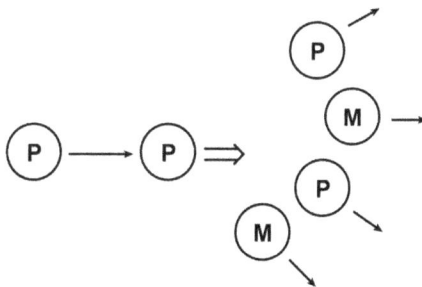

Figure 63. Inelastic collision between protons.
New particles are formed.

Proton-Electron Collision

The Stanford particle accelerator (SLAC) was built in the 1960s in order to accelerate electrons and have them collide with protons. This was the first accelerator where electrons were sufficiently energetic, and their wavelengths sufficiently short, so as to hit a single quark inside a proton. As years went by even more powerful accelerators were built, and the energy of particle collisions grew even higher. One of the more interesting phenomena discovered in those experiment was that these high-energy collisions were almost always inelastic, and, as energy levels grew, the rate of elastic collisions became ever smaller. At very high energies it was found that only one-thousandth of all collisions were elastic. This means that the electron almost always succeeded in making the quark "split" and produced another pair of particles. This is an example where energy is transformed into matter.

The alternative, which almost never occurred, was that a single quark of the proton absorbed the energy of the collision and accelerated the entire proton. But a lone quark that suffered a strong impact by the electron would find that very hard to do.

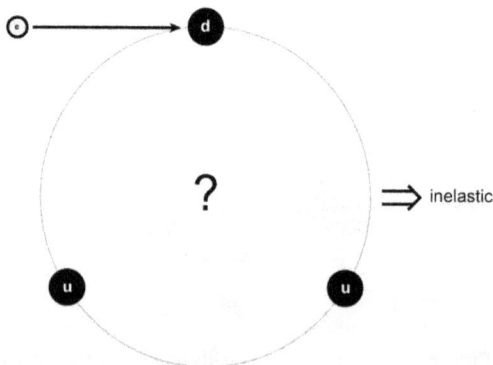

Figure 64. When a high-energy electron hits a quark inside the proton, it almost always (in 99.9% of cases) results in inelastic collision and the formation of new particles.

Let's recall the alternative models that I mentioned earlier in this book. Were the proton to conform to the D model—namely, were it to have a massive core with no electric charge that pulls three quarks that surround it. Since the core is a chargeless elementary particle, and, because it lacks a charge, the electron never interacts with it. Therefore the D model is compatible with this phenomenon.

It would be interesting to compare the E model, which describes additional internal quark shells, which are analogous to a multishell atom. Here the internal quarks are attracted to the core in a much stronger interaction than that of the outer quarks. If this model were true, it would have been harder to produce an electron beam that is energetic enough to penetrate the massive core inside the proton, and so this model also explains why the probability of elastic collision is so small.

QCD easily explains why the high-energy collision between electrons and protons is almost always inelastic: according to QCD, the proton has no massive core, and the collision always takes place between an electron and a quark, which almost always causes them to produce new particles, and leads to inelastic collision.

Proton-Proton Collision

In the case of energetic proton-proton collisions, particle accelerators are able to provide protons with much higher energy than that of electron accelerators, and therefore the wavelength of the proton's components is shorter. This allows a much deeper penetration into the colliding protons.

If QCD is true, then protons are made of quarks that cannot suffer a strong impact without "splitting," and, therefore, according to QCD, even proton-proton collisions should almost always be inelastic.

Experimental results, however, tell an entirely different story.

It turns out that when sufficiently energetic collisions take place, approximately 15% of all collisions are elastic! This is a huge percentage, and it is about one hundred times greater than the rate of elastic collisions between protons and electrons at similar energies. How can this be?

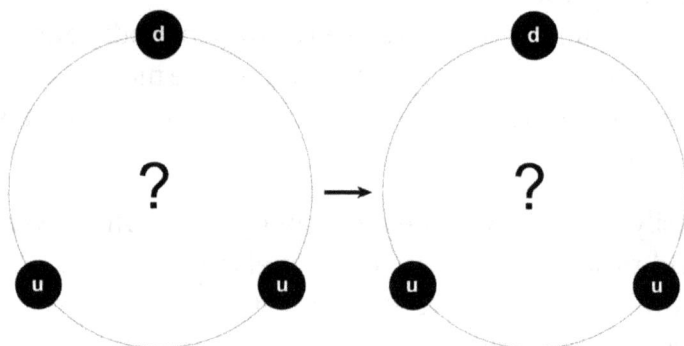

Figure 65. 15% of all energetic proton-proton collisions are elastic.

QCD does not explain this phenomenon, nor can it, as according to QCD the collision takes place between the quarks of one proton and those of another. Under such circumstances the number of elastic collisions ought to be even smaller, since the energy that characterizes proton-proton collisions is even higher, and the quarks are therefore unable to suffer a strong impact and still allow an elastic collision.

In contrast, the D and E models do explain this phenomenon: an elastic collision between energetic protons can form as a result of the collision between the massive cores of the protons. The cores hit each other just like two billiard balls, and that explains the higher frequency of elastic proton-proton collisions.

The fact that QCD cannot explain the results of proton-proton energetic collisions was mentioned in academic literature by the Russian physicist A. A. Arkhipov over a decade ago.[129]

Other physicists apparently prefer to ignore this problem.

---

129 A. A. Arkhipov, *On global structure of hadronic total cross sections,* 1999, arxiv.org/PS_cache/hep-ph/pdf/9911/9911533v2.pdf.

Deeper Inside

In the early 2000s scientists were able to collide proton and antiproton beams using the Tevatron particle accelerator in Illinois at hitherto unseen energy. In order to understand the significance of their results, we will examine several similar experiments conducted over the last hundred years.

An Inspiring Experiment

In 1913 physicists James Franck and Gustav Hertz conducted one of the first experiments that showed that the energy levels of electrons inside the atom are never random, and actually have very specific values. The results of that experiment supported the model prevalent at the time—namely, Bohr's model. Bohr's model was later replaced by the second quantum theory, but the experiment did provide a source of inspiration for other experiments.

Franck and Hertz were able to produce an electron beam that traveled inside a tube that contained argon gas. The beam was accelerated by an electric field. They were able to control the strength of the electric field and were thus able to control the energy of the electron beam. The electrons traveled from one side of the tube to the other. Franck and Hertz measured the intensity of the electric current produced by the electrons arriving at the other side of the tube.

Figure 66 describes the intensity of the current produced at the other end of the tube as a function of electron acceleration. At first we see that the current intensified with the increase in electron energy, and then it suddenly dropped, only to rise again, and then drop again, and so on.

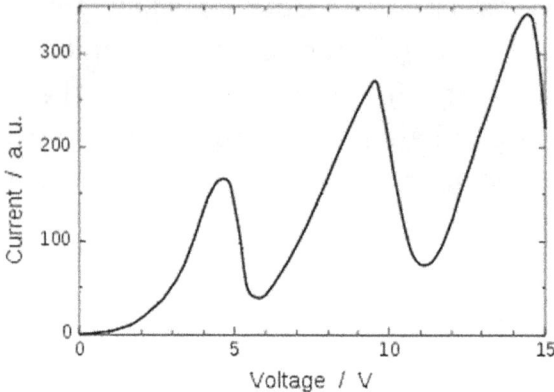

Figure 66. Current intensity on the other
side of the tube (Franck-Hertz experiment).

How do we explain these results? At low energy, electrons are unable to excite the argon atoms, and thus continue to travel to the other side of the tube without losing any energy. At this point current intensifies as the electrons are more and more accelerated. When the electrons reach a certain threshold that allows them to excite the argon atom, the intensity of the current suddenly drops because the electrons lose some of their energy, which is transferred to the argon atom. Later, the intensity of the current continues to increase, and drops again later when the electrons are sufficiently energized to hit the argon atoms twice and excite two atoms, and so on.

This experiment inspired many other experiments, including some that are being carried out using today's particle accelerators.

The Cross Section of Multi-Shell Particles

As I mentioned in the previous chapter, when dealing with electron scattering, the cross section, which is proportional to the number of collisions between a beam and its target, falls naturally with an increase in the beam's energy. This is true when a beam hits elementary particles. When it hits complex particles, the situation is more complicated.

For example, let's consider the calcium 40 nucleus. In this nucleus there are twenty protons and twenty neutrons. As we already know, these nucleons are arranged in shells. Let's examine the cross section graph of an electron beam that collides with these calcium nuclei.

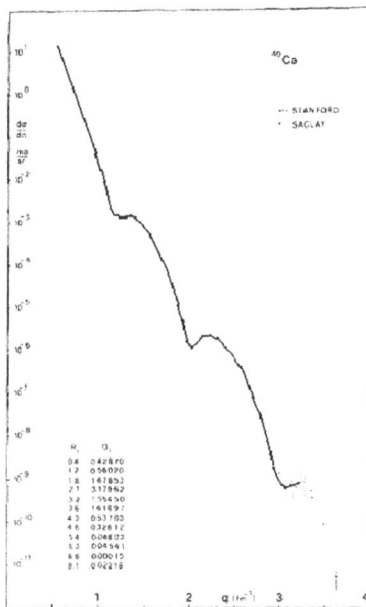

Figure 67. Cross section graph depicting the collisions of an electron beam and a calcium 40 nucleus.[130]

---

130   I. Sick *et al.*, *Charge density of* ^{40}Ca, Physics Letters B, Volume **88**, Issues 3–4, (1979). p. 245–248.

The graph falls as electron energy increases, as one would expect from the interaction between an electron and charged particles. However, every now and then we find a small rise before the graph continues to fall.

Why does this happen? The cross section graph of a low-energy electron stems from the interaction between the electron and the nucleons situated in the calcium nucleus's outer shell. When the electrons are energetic enough, they manage to penetrate the outer shell and interact with a more internal shell. And so, each such rise found in the cross section graph stems from the fact that other shells begin to interact with the electrons (see Figure 68).

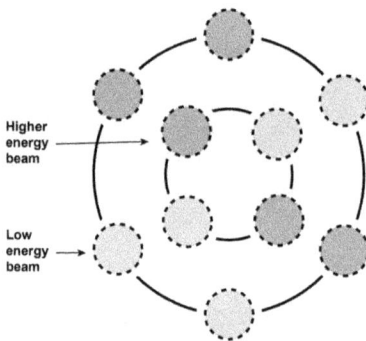

Figure 68. The interaction between a particle beam
and an atomic nucleus.

The Proton's Cross Section

From our previous discussion we learned that a cross section graph provides an indication about the inner structure of complex particles. Therefore, the cross section of the interaction between protons and another particle beam (electrons or protons) can teach us about the inner structure of the proton.

Since the 1970s, when scientists were able to use stronger electron beams capable of hitting a single quark, the cross section graph has been steadily falling with the increase in collision energy. During the 1980s this fall provided evidence that protons and neutrons do not contain any other massive particles. Scientists have been using the falling cross section graph as a supporting argument for QCD, according to which quarks are the only massive particles inside the proton.

But in the 1990s, in an experiment conducted at the DESY laboratories in Germany, the electrons reached an unprecedented energy. To the scientists' astonishment, the amount of incidents observed was much greater than anticipated. The experiments were conducted by two independent groups who reached the same results. The two groups published two articles that describe their work.[131,132]

After the scientific findings were published in a regular physics journal, without being interpreted, two of the experimenters, Frank Sciulli and Steven Ritz, were interviewed by Colombia University News.[133] Sciulli stated, "If the results are not a statistical fluke, new physics has been observed. One possibility is that our understanding of what's inside the proton is somehow wrong."

---

131   C. Adloff *et al.*, *Observation of Events at Very High Q^2 in ep Collisions at. HERA*, Z. Phys **C74**, 191 (1997).

132   J. Breitweg *et al.*, *Comparison of ZEUS Data with Standard Model Predictions for ep -> eX Scattering at High x and Q^2*, Z. Phys **C74**, 207 (1997).

133   Columbia University News, 1997. www.columbia.edu/cu/pr/97/19058.html.

Ritz added that the probability of a statistical fluke was less than one percent.

Were they exaggerating when they said that? Not at all. The only explanation for a rise on the cross section graph is the existence of other massive objects that carry an electric charge inside the proton. Pure and simple. And this explanation contradicts the Standard Model.

When I asked a physicist who has published hundreds of articles on the subject, he came back to me and said that this was a classic instance of scientists who want to "beat their own drums." Sciulli and Ritz are two members of a group that included hundreds of scientists who are mentioned in the original articles.

I was surprised by that blunt, rather aggressive reply. When I called Frank Sciulli more than ten years after the article was published, he told me that he *does* think that there was a statistical fluke.

But in the 2000s particle accelerators were able to reach energy levels that removed any shadow of a doubt that may have been cast on these results.

Fermilab, Illinois, the 2000s

Using the Tevatron particle accelerator at Fermilab in Illinois, scientists were able to cause proton beams to collide with proton or antiproton beams at hitherto unseen energy. Scientists saw a clear rise in the cross section graph of proton-proton collisions. This rise is found in Segment 5 of the Figure 61 cross section graph described in an earlier chapter, and which appears again in Figure 69 here.

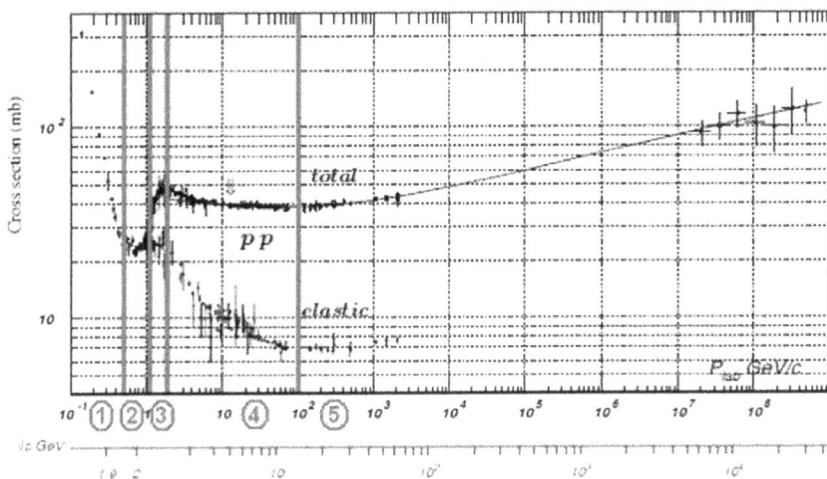

Figure 69. Cross section graph of inter-proton collisions,
divided into five segments.

In 1997, when the cross section graph of electron-proton collisions turned out to be bigger than expected, Frank Sciulli came out and said that if the results were not a statistical fluke, then we have failed to understand the structure of the proton. And indeed we find the results of the experiment that measured the collision of two protons at higher energy are found in the graph in Figure 69. In Segment 5 there is a clear rise of the elastic cross section that corresponds to the increase in collision energy. Is it reasonable to expect that the confirmation of findings that indicate the existence of massive particles inside the proton will lead to a fierce debate about the veracity of the Standard Model?

Well, not really. The graph does appear in textbooks, but the obvious implication of these findings is never mentioned, and scientists are notably silent about it. The rise of the cross section that depicts inter-proton collisions can only be explained by the existence of additional massive particles inside the proton, an explanation that would contradict QCD.

When I asked a Nobel Prize laureate about this finding, and how we can reconcile it with QCD, he preferred to ignore the question. Another notable scientist told me that he "heard that they're trying to explain the phenomenon using pomerons."

This reply can teach us that some physicists are troubled by these findings. Pomerons are hypothetical massive, chargeless particles, and the Ukrainian scientist Pomeranchuk postulated that they could affect proton interactions back in the 1950s, even before QCD was developed. The pomerons were never found.

So what have we seen here? Here's a recap:

- According to the laws of physics, the large number of proton-proton elastic collisions—and particularly the increase in the number of those collisions with an increase in energy—necessitates the existence of additional massive objects inside protons apart from quarks.

- According to QCD, no such objects exist.

- Therefore, there is a contradiction between QCD and the large number of elastic collisions.

- In order to solve this contradiction, some particle scientists try to adopt a theory invented over a decade before QCD was born—a theory that contradicts QCD—and combine it with QCD. This theory, which is adopted in order to "save" QCD, is based on particles that do not exist.

- The rest simply ignore the contradiction.

It's amusing, or maybe amazing, how an entire scientific community can avoid drawing conclusions from a finding that so blatantly contradicts QCD.

Out of the alternative models we mentioned, the D model fails to explain the phenomenon, whereas the C and E models, according to which inner quark shells exist, do explain the rise of the cross section graph.

The Mass of a Proton's Three Quarks

Until now we saw that, of all the alternative models we proposed, the D model and particularly the E model are in fact two variants of the liquid drop model, and they are surprisingly able to explain the phenomena that challenge QCD. I assume that readers might at this point wonder how it is possible that these two models, which argue for the existence of a massive core inside the proton and neutron, can compete with a theory that claims that it doesn't exist. A simple calculation of masses should easily refute one of the approaches.

It turns out that the mathematics behind such an endeavor is not that simple. Let's begin with the most obvious example. A pion contains a u quark and a $\bar{d}$ antiquark, and has a mass of 140 MeV, whereas a proton containing three quarks has a mass of 938 MeV. It appears that this finding alone refutes QCD, as a proton should supposedly be 50% more massive than a pion, and because the mass of a proton is almost seven times greater than that of a pion, it follows that QCD is false.

But this line of reasoning is wrong. The calculation of hadron masses (mesons and baryons) is more elaborate, and it is very dependent on other factors beyond the simple addition of hadron masses.

I wrote this chapter for those who want to better understand the calculations of hadron masses. If you are not interested in calculations, please feel free to skip this chapter.

———

195

The Mass of Bound Particles

Let's recall that mass and energy are actually two sides of the same coin. If, for example, we have two bound particles, then we need to invest energy in order to separate them. As energy is a form of mass, then after we invest that energy and create a separation, we will obtain two particles, and the sum of their masses will be equal to the mass of the bound particle plus the energy invested in order to separate them.

This energy is known as binding energy. The nucleus of an ordinary helium atom, for example, is composed of two protons and two neutrons, and it is about 1% smaller than the sum of the masses of these two protons and two neutrons. This kind of disparity between the masses is the source of the huge amount of energy produced by nuclear bombs and reactors.

Let's examine another case of mass changes that stem from changes in energy. If we take an electron inside the hydrogen atom that is situated at the lowest energy level—that is, 1s—then its energy is smaller than that of an electron situated at a higher energy level—say, 2p. But the difference in mass we find between these two states is negligible: about 1/50,000.

In quark-bound particles, however, things are different, and the change found in mass is a lot more significant.

Let's consider, for example, three mesons, all of which are composed of a u quark and a $\bar{d}$ antiquark. There are other mesons composed of these quarks, but the three we mention here are presented as an example. We would supposedly expect them to have a similar mass, but that's not actually the case:

$$M(\pi^+) = 139 \text{ MeV}$$
$$M(\rho^+) = 770 \text{ MeV}$$
$$M(\rho^+_3) = 1690 \text{ MeV}$$

It turns out that three mesons, each of which is composed of the same quarks, possess completely different masses, to the point that one is more than ten times bigger than another! As is the case for the hydrogen atom, the meson is also composed of two bound particles, which can be situated at different energy levels. In the case of mesons, the changes in energy levels can bring about a very significant change in mass.

This situation is very different from what we know about electric forces, because in the case of strong forces most of the quark mass disappears when they are bound.

The Mass of Quarks inside Protons and Pions

As I mentioned earlier, the mass of a pion composed of two quarks is much smaller than that of a proton composed of three quarks. The quarks inside pions are situated at the lowest energy level, as there are no lighter mesons. The quarks inside protons are also situated at the lowest energy level, as there is no lighter baryon. Therefore, one would expect to find quarks inside a proton and those inside a pion to have more or less the same mass. This statement is based on the assumption that the strength of the interaction between two quarks is not dramatically different from the interaction between a quark and an antiquark.

QCD provides a whole theory that explains why the mass of pions is so small. We will make no attempt to confront that theory. In this chapter we will use standard physics and see if the C, D, and E models are able to explain the differences in mass between pions and protons.

On the face of it, it appears that all three models naturally explain the differences in mass, as they include additional quarks or a massive core that can explain why a proton is much heavier than a pion.

But there is another finding we need to take into consideration. In the 1970s, scientists measured the total mass of quarks inside a proton and found it equals about one-half of the proton's mass. By the way, this mass also includes the other quark-antiquark pairs, but for the purposes of this discussion we will ignore them in order to simplify our considerations.

The mass of quarks equals about half of the proton's mass—namely, about 470 MeV. Therefore, the mass of each proton-bound quark is approximately 155 MeV. We're assuming here that the mass of the u quark is almost the same as that of the d quark, a rather reasonable assumption according to many findings, and one that is accepted by all.

In a pion, however, the mass of a single quark is about 70 MeV—that is, less than half of the proton-bound quark. How can this be?

This finding seems to contradict the D model, which assumes that a proton has a massive nucleus that binds three quarks, and that no other quarks exist. The C and E models explain this finding with ease: the quarks inside a proton are not situated in its lowest shell, and therefore they are at a higher energy level when compared to the internal quarks, which are usually 1s particles. In a pion, however, the quarks are usually 1s particles, and therefore their energy level is lower, and their mass is therefore smaller.

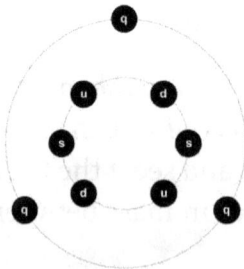

Figure 70. C model: Quarks in the outermost shell have much more energy than internal quarks.

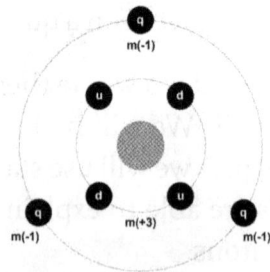

Figure 71. E model: Quarks in the outermost shell have much more energy than internal quarks.

The Proton Spin Crisis

In this chapter I discuss the proton spin crisis mentioned several times earlier in this book. I remind readers here again that the assumption made in the sixties was that the *uud* quarks were the only massive particles that exist inside a proton, and that they were s-wave particles of the lowest energy level.

Let's also recall that an s-wave particle does not possess orbital angular momentum. By the way, this assumption is not entirely true when it comes to particles like quarks, which travel almost at the speed of light, but let's ignore that for now.

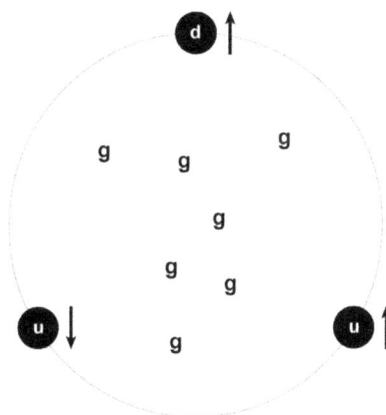

Figure 72. The structure of a proton according to the founders of QCD: two quarks with spins that are parallel to the proton's, and one quark with opposite spin. None of the quarks possess orbital angular momentum.

The proton is a spin ½ particle, and since according to the founders of QCD quarks do not possess orbital angular momentum, the sum of their spins ought to be identical to that of the proton. From here it follows that two quarks ought to have parallel spins, and one quark ought to have the opposite spin. In this arrangement the three quarks' spin adds up to ½, which is the same as the proton's spin.

This was a commonly held belief at the time. In the eighties an experiment was conducted in order to confirm that this was indeed an accurate depiction of the proton.

The Experiment

A group of scientists known as the EMC (European Muon Collaboration) conducted an experiment whereby it bombarded polarized liquid hydrogen with polarized muon beams. Liquid hydrogen is made mostly of protons, and when polarized all of its protons possess parallel spins. In its natural, nonpolarized state, the hydrogen contains a "mixture" of spins—namely, the protons' spin can face any direction permitted by the laws of physics. Polarized hydrogen is obtained by the operation of a very strong magnetic field that is used to force the protons' spin to take the direction of the field.

Muons are similar to electrons, and they too were polarized so that their spins were facing the desired direction. During the experiment, muon beams were energetic enough so that when a muon hit a proton, it actually collided with a single quark inside the proton.

What did the scientists expect? They expected, as the simple model I mentioned earlier indicates, that two-thirds of all muon-quark collisions would result in their having parallel spins, and one-third of them in opposite spins.

The Crisis

The results of the experiment[134] astounded the scientists.[135] It turned out that the number of times a muon collided with a quark and had a parallel spin was almost similar to the number of times a collision resulted in the opposite.

Many scientists, some of who are still convinced that a proton's *uud* quarks ought to be s-wave particles, have explained the data by saying that gluons inside the proton arrange themselves asymmetrically in a way that produces spin; in other words, they are the ones that carry most of the proton's spin. Even if we did believe in the existence of gluons, such behavior by gluons is unlike any other model that describes bound particles. In the atom, for example, one may assume that the internal shells are symmetrical, and that their contribution to the atom's total spin is zero, and therefore only the electrons in the outer shell contribute to the atom's total spin. However, scientists were required to formulate models in which the proton's spin was carried by gluons.

That, in principle, is the proton spin crisis, one of the most important mysteries that persists to this day.

---

134   J. Ashman *et al, A measurement of the spin asymmetry and determination of the structure function g1 in deep inelastic muon-proton scattering*, Physics Letters B, Vol **206**, (1988). p. 364–370.

135   Ivars Peterson, Science News 1997. *"In 1988, however, physicists were shocked to find experimental evidence suggesting that very little–perhaps none–of the proton's spin comes from the spin of the quarks..."*

Orbital Angular Momentum

In fact, the A, C, D, and E models can all resolve the spin crisis if we once again apply the configurations concept, which accurately describes atoms and can also be used to describe the quark state of protons and neutrons.

We will start with the A model, the one espoused by QCD, and assume that configurations exist as they should in every particle made up of a combination of three or more components. As previously mentioned, even the seemingly simple ground state of the helium atom, which is composed of a nucleus and two electrons, is characterized by configurations,[136] and even there one finds electrons that are considerably less likely to be s-wave particles.[137] The proton also has quark-antiquark pairs, and this addition increases the number of particles and significant configurations.

The calculation of states composed of multiple configurations is far from simple, and they employ Wigner-Racah algebra. However, it's enough to understand that the existence of many configurations adds states with an orbital angular momentum that is greater than zero. This substantially reduces the contribution of a quark's spin to the proton's total angular momentum.

The D model resolves the crisis in a similar fashion.

The C and E models resolve the spin crisis with even greater ease, since according to them quarks are located in the outer shell, and they are added by other quarks situated inside closed internal shells. In this scenario, we could find p-wave external quarks, as

---

136   G.R. Taylor and R.G. Parr, *Superposition of configurations: The helium atom*, Proc., Natl.Acad. Sci. USA **38**, (1952). p. 154–160.

137   AW Weiss, *Configuration Interaction in Simple Atomic Systems*, Phys. Rev. **122**, (1961). p. 1826–1836.

is the case for electrons inside atoms that possess more than four electrons. Here we have a much larger number of significant configurations.

There is another reason for the fact that quarks must possess angular momentum—namely, the Dirac equation, which describes quarks. A solution of this equation includes an "upper part" and a "lower part." If the spatial angular momentum of the upper part vanishes, which means that it is an s-wave, then the lower part has angular momentum 1 and it is a p-wave. Since inside the proton quarks move at almost light speed, the lower part of the solution provides a substantial contribution, and therefore the angular momentum of the quarks is significant for this reason as well. But according to calculations, it is not enough to explain the entire effect that originates from the proton spin crisis.[138]

In fact, an accurate calculation of configurations can show the number of cases where the quark's spin is parallel to that of the proton. Today, however, there exists no conventional theory that tells us how to perform that calculation.

Recognizing the Existence of Orbital Angular Momentum

In experiments conducted in the 2000s at the Jefferson laboratory in Virginia, it was found that a quark inside a neutron sometimes possesses most of the neutron's energy, and that its spin is op-

---

138 A.W. Thomas, *Interplay of Spin and Orbital Angular Momentum in the Proton*. Physical Review Letters, **101**, 102003 (2008).

posite to that of the neutron.[139] According to QCD's naïve perception, as described in the beginning of this chapter, such a scenario is impossible.[140]

One of the scientists who participated in the experiment, Xiangdong Ji, adds his interpretation: "It might also indicate that orbital motion of particles within neutrons, in addition to those particles' spins, are more important than previously recognized."[141], His fellow researcher Xiaochao Zheng points out: "Given that neutrons and protons are sister particles, called nucleons, the new findings apply to both."

Will scientists recognize the fact that proton-bound quarks carry orbital angular momentum? If so, would they admit that the two underlying assumptions on which QCD was founded are wrong? These assumptions are:

- Quarks carry all of the proton's mass.

- They have no orbital angular momentum.

These assumptions are the reason physical theories known in the sixties were considered unable to describe protons, neutrons, or similar particles such as Δ^{++} or Ω^-. These assumptions were crucial for the invention of QCD, a theory that relies on premises hitherto unseen in physics, and that, as we've seen so far, fails again and again.

---

139   Peter Weiss, *Topsy Turvy: In neutrons and protons, quarks take wrong turns*, Science News, vol 165, 2004.

140   Ibid. *"'That's very disturbing,' comments theoretical physicist Xiangdong Ji of the University of Maryland at College Park."*

141   Ibid. *"'The finding suggests that scientists may have erred in calculations using fundamental theory to predict quark behavior within neutrons,' he says."*

In a lecture given by the physicist Stanly Brodsky in 1989, he mentions five phenomena that contradict QCD predictions, two of which are referred to in this book—the proton spin crisis and Krisch's experiments with polarized protons. Later in his lecture he states that these phenomena do not necessarily indicate QCD's failure, but rather the fact that the structure of baryons is a lot more complicated than they had previously thought.[142]

Does Brodsky's argument make sense? The whole premise of QCD was a simplistic idea about the structure of baryons, which led to us being unable to understand the properties of the baryons Δ^{++} and Ω^-. Had it been known in the sixties that baryons were a lot more complex than hitherto imagined by scientists, then the QCD theory probably would never have seen the light of day.

---

142 Stanley J. Brodsky, *Color Transparency And The Structure Of The Proton In Quantum Chromodynamics*, SLAC, June 1989. "*...All of these anomalies suggest that the proton itself is a much more complex...the apparent discrepancies with experiment are not so much a failure of QCD, but rather symptoms of the complexity and richness of the theory...*"

Baryon Number Conservation Law

Protons and neutrons are examples of baryons. In all experiments conducted thus far the number of baryons involved in any observed process is conserved, including processes that involve modern particle accelerators. This means that if we start a process with a certain number of protons and neutrons, then we will eventually have the same number of protons and neutrons at the end of it. This is known as the baryon number conservation law.[143]

In this calculation, the antibaryon particle has a baryon number of -1. This means that formation of a baryon-antibaryon pair does not violate this law, as the sum of the two equals zero.

QCD is neutral about this law. According to QCD, it's possible to have processes that violate the baryon number conservation law. As a result, quite a few theories were developed that try to violate the baryon number conservation law. Such theories do not form a part of QCD, but they do not contradict it, either. They belong to a collection of theories that are "beyond the Standard Model."

Proton Decay

The proton is the lightest type of baryon. If a free proton spontaneously decays into other particles, they should not include any

---

143 Eidelman *et al.* (Particle Data Group), Phys. Lett. **B592**, 1 (2004).
     "*No baryon number violating processes have yet been observed.*"

baryons. In the seventies Georgi and Glashow[144] postulated that the proton indeed decays into a pion (π^o) and a positron.

The question of whether a proton actually decays is considered one of the most important open questions in physics.[145]

The Baryon-Lepton Difference Conservation Law

Most "beyond the Standard Model theories" argue that it's possible to violate the baryon number conservation law, but the difference between the number of baryons and leptons is conserved. Leptons are the three types of electrons (electron, muon, and tau) and the three types of neutrinos. Georgi and Glashow's theory also conserves this difference because the positron has a lepton number of -1 since it is an antilepton.

Even though we have never observed a process that violates the number of baryons or leptons, the Standard Model is still facing a problem, because according to the model the neutrino should be massless. However, the neutrino's mass has been measured, and, while very small, it is not equal to zero. Some theories explain why

144   Howard Georgi and Sheldon Glashow, *Unity of All Elementary-Particle Forces*, Physical Review Letters, **32** (1974) 438.

145   en.wikipedia.org/wiki/List_of_unsolved_problems_in_physics (2013).

neutrinos have mass,[146,147,148,149,150] and according to them only the difference between the number of baryons and leptons is conserved.

An Alternative Explanation for the Baryon Number Conservation Law

If we accept the C, D, or E model, which postulate a massive core inside the baryon, then the baryon number conservation law seems self-evident. Because the energies produced by our particle accelerators can only affect the outermost quarks of the baryon, that leaves us with its inner core. In fact, the number of baryons is always equal to the number of cores, and the three outermost quarks, while they do change themselves, never affect the baryonic core.

This is a bit like a chemical reaction that involves a reaction between atoms, ions, or molecules. It is a well-known fact that a chemical reaction only affects the outermost shell of the atom,

146 P. Minkowski, $\mu \rightarrow e\, \gamma$ *at a Rate of One Out of 1-Billion Muon Decays?* Physics Letters **B67** (1977).

147 M. Gell-Mann, P. Ramond and R. Slansky, in Supergravity, ed. by D. Freedman and P. Van Nieuwenhuizen, North Holland, Amsterdam (1979). p. 315–321.

148 T. Yanagida, *Horizontal Symmetry and Masses of Neutrinos.* Progress of Theoretical Physics **64** (1980).

149 R. N. Mohapatra, G. Senjanovic, *Neutrino Mass and Spontaneous Parity Nonconservation.* Phys. Rev. Lett. **44** (1980).

150 J. Schechter, J.W.F. Valle; Valle, J., *Neutrino masses in SU(2) $\otimes$ U(1) theories.* Phys. Rev. **22** (1980).

and, because the atomic nucleus does not participate in chemical processes, the number of atoms will always be conserved.

If we assume for a moment that a baryon core is analogous to an atomic nucleus, and that quarks are analogous to electrons, then the baryon number conservation law becomes perfectly clear.

A Summary of
Unexplained Phenomena Thus Far

We will now present a list of all the problematic phenomena that indicate the existence of massive particles inside the proton.

Table 7. Problematic phenomena associated
with massive particles inside the proton.

Contradicting Phenomena	Suspicious Phenomena
13. A substantial number of proton-proton collisions are elastic.	3. Baryon number conservation law.
14. The proton-proton collision cross section graph rises at very high energies.	4. Protons never decay.
15. The proton spin crisis.	
16. Protons' quarks carry orbital angular momentum.	

ATTRACTION OR REPULSION?

The Quark and Antiquark
Inside Protons and Neutrons

According to QCD, quarks attract one another, antiquarks attract one another, and quarks attract antiquarks. They all attract each other. In the following chapters we will examine whether this assumption conforms to experimental data.

A surprising conclusion drawn from quantum physics is that in a hydrogen atom, which is made up of a proton and an electron, we occasionally find that an additional electron-positron pair is formed. The electron and positron immediately annihilate one another, and measurements indicate that about one-millionth of the time (or in one-millionth of all cases where we measure a hydrogen atom) we find another such pair. The existence of this pair explains an effect known as the Lamb shift, which was discovered in the 1940s.

According to quantum field theory, particle and antiparticle pairs exist in every quantum state, in both the proton and the neutron (and mesons, naturally). Therefore, quark and antiquark pairs form inside the proton and neutron, and these in turn annihilate themselves only to form again later. In nucleons this phenomenon is much more common than it is inside the hydrogen atom, and quark-antiquark pairs have been measured directly. Such pairs exist "half the time."[151]

Where Is the Antiquark Normally Found?

In this paragraph I use several new terms, and if you are not a physicist, please forget them immediately after reading, and remember only the bottom line. One of the most useful measures when analyzing proton bombardment events is called Bjorken-x, after the physicist James Bjorken. It turns out that the antiquark has a Bjorken-x that usually exists in a relatively narrow region. For "physics jargon fans" I will mention that this implies that they have a weaker Fermi motion. This fact, together with a very basic quantum theoretical property known as the uncertainty principle, indicates that antiquarks tends to be located in the outer section of the nucleon, whereas quarks tend to be found in the inner section.[152]

The nature of quarks and antiquarks according to Bjorken-x is reported in Perkins's book without any added interpretation, and has been known to us for decades. On the face of it, it seems to contradict QCD. The reason for that is that according to QCD, quarks and antiquarks also attract each other. Since the antiquark is being pulled by four other different quarks, it should be found in a volume in space that is not larger than the volume that contains all the quarks, and therefore this fact contradicts QCD.

If we look at the alternative models, we see that the D and E models postulate a center with a "positive strong charge," similar to the

---

151   D. H. Perkins, *Introduction to High Energy Physics*, (Addison-Wesley, Menlo Park, CA, 1987). p. 281.

152   Ibid.

antiquark, which also possesses a positive strong charge. Now, if the strong force is analogous to the electromagnetic force then such a force explains why the antiquark is pushed to the edges of the proton (see Figure 73).

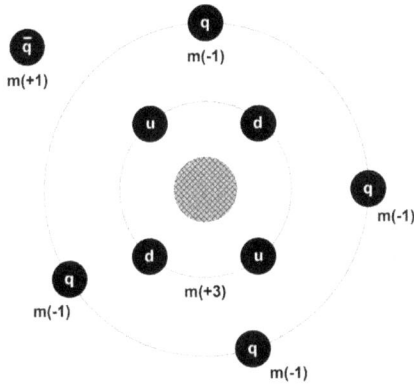

Figure 73. The antiquark is being pushed away from the positive center of the nucleon according to the E model. The letter m denotes a strong charge. The sum of the core's strong charges along with its internal shell of the quarks is +3. Each quark has a negative strong charge of -1, and the antiquark has a positive strong charge of +1. In the outermost shell we find four quarks, three quarks of the nucleon and the counterpart of the antiquark.

The E model has an advantage over the D model, since it more easily explains why the volume of the proton is larger than that of the pion. Had all the proton's quarks been located inside the internal shell, its volume would have been smaller than that of the pion. This is similar to the helium atom, which is actually smaller in volume than hydrogen. According to the E model, the volume of the proton is larger than that of the pion because the quarks are situated in the outermost shell and at least one internal shell exists. The situation in atoms is similar: multi-shell atoms are greater in volume than hydrogen or helium atoms.

If we take into account the fact that quarks are supposed to arrange themselves in multiple configurations, it therefore follows that the D model may also explain why the proton's volume is bigger than the pion's, which has only one configuration (if we ignore the fact that it also contains quark-antiquark pairs).

The Distribution of Charge in Neutrons

An important quantity that helps us understand the structure of neutrons and protons has to do with the electric charge found in the volume that contains the nucleon. This quantity is obtained statistically by calculating an average number that spans many events of proton and neutron bombardment by electrons, or muons.

Measurements have shown that the negative charge tends to accumulate in the outermost section of the neutron, whereas the positive charge tends to be found nearer to its center.[153,154]

The immediate explanation for this phenomenon is actually a combination of two known phenomena. As readers may remember, the neutron contains an additional quark pair that exists about 50% of the time. What quark pair are we talking about? Is it $u\bar{u}$ or $d\bar{d}$? The neutron is characterized by udd quarks, and is known to contain a $u\bar{u}$ pair more often than it contains a $d\bar{d}$ pair.[155] Therefore, the neutron contains a $\bar{u}$ quark at a relatively

153   K. Nakamura *et al.* (Particle Data Group), J. Phys. G **37**, 075021 (2010). pdg.lbl.gov/2010/listings/rpp2010-list-n.pdf, p. 4.

154   C. Amsler *et al.* (Particle Data Group), Phys. Lett. **B667**, 1 (2008).

155   M. Alberg, *Parton distributions in hadrons*, Prog. Part. Nucl. Phys., **61**, 140 (2008).

high frequency. If we also recall the phenomenon indicated by the data included in Perkins's textbook, we would find that the antiquark is located at the periphery of the neutron.

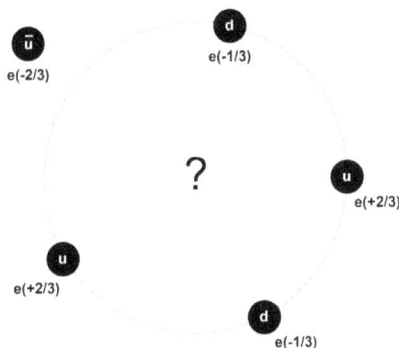

Figure 74. A simplistic illustration showing why the negative electric charge of the neutron is found in its periphery. The $\bar{u}$ antiquark has a negative charge of -2/3, and is situated in the neutron's periphery.

The problem is that particle scientists disregard this phenomenon when it comes to the position of the antiquark in the nucleon's periphery, despite the fact that we have been aware of its existence since the 1970s. Maybe they ignore this fact because it contradicts QCD. Therefore, in the early 1980s scientists developed a model called the cloudy bag model, or CBM, which was meant to explain this phenomenon. According to this model, a cloud of negative π⁻ surrounds the neutron, thus forcing the negative charge to concentrate in the neutron's periphery.[156,157]

156 A. W. Thomas *et al.*, *Cloudy bag model of the nucleon*, Phys. Rev. D24, 216 (1981).

157 A.W. Thomas, *The cloudy bag model*, Czechoslovak Journal of Physics B 1982, Volume 32, Issue 3, p. 239–247.

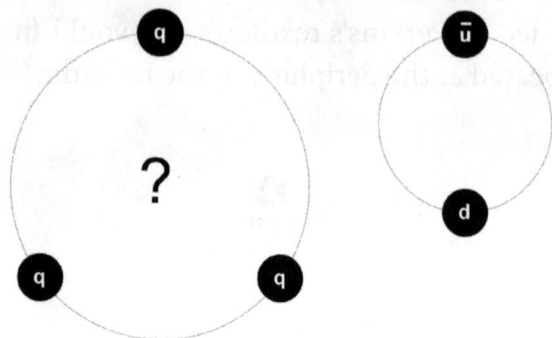

Figure 75. A neutron according to the cloud bag model.
The cloud is made up of π⁻ particles.

Not only does QCD fail to explain why pion clouds exist, but it is also extremely unlikely that such pion clouds are even possible, owing to several simple and well-known principles. We should remember that according to CBM the pions are situated in the edges of the bag, and so the forces operating on them should be similar to the nuclear force. The idea is very unlikely because the nuclear force is negligible compared to the pion's mass.

But there is another argument that makes the CBM idea even less likely. We must remember that the strong force abides by parity conservation. Indeed, in processes governed by the strong interactions, parity conservation is a phenomenon that has been measured with an accuracy of almost ten decimal places. Therefore, because parity is conserved, and because pions are odd-parity particles whereas nucleons are even particles, it then follows that as soon as the pion leaves the nucleon its parity is immediately affected, unless the nucleon's parity itself is reversed and it transforms into an odd particle as well. Such parity reversal requires a substantial amount of energy and it is unlikely to occur in neutrons. Alternatively, the pion cloud must contain an even number of pions. Such an arrangement would also greatly increase the amount of energy and reduce the feasibility of the model.

Another problem of CBM relates to the preference of negative pions over positive pions. The origin of these effects is unclear, but they are crucial for the relevance of CBM as an explanation for the position of negative charges in the outermost regions of the neutron.

The Distribution of Charges in Protons

A similar tale can be told about the proton. The proton contains *uud* quarks and has more $d\bar{d}$ pairs. Since the $\bar{d}$ antiquark is located in the proton's periphery, and its charge is +⅓, it follows that part of the positive charge should be pushed into the outer regions of the proton.

Indeed, this phenomenon was discovered in the early 2000s, leaving physicists astounded.[158]

This phenomenon is less tangible than that which takes place in neutrons for two reasons: one is that the charge of a $\bar{d}$ antiquark is only ⅓, whereas in the neutron the $\bar{u}$ quark's charge is --⅔. The second reason is that the proton contains a positive charge in any event, and a concentration of some of the charge in the periphery slightly more than at the center is difficult to measure. For this reason, the charge distribution in protons was finally measured

---

158 Peter Weiss, *New probe reveals unfamiliar inner proton*, Science News, vol **159**, May 5, 2001. *"The new findings come as a shock to nuclear and particle physicists who have considered this aspect of the proton to be well understood for more than a half century."*

only in the 2000s, whereas in the neutron's case it was measured even before QCD came into existence.

It turns out that the existence of positive charge in the periphery of the proton—which to us seems natural and a result of known facts that have been available for decades—was a big surprise for particle physicists in the 2000s. Here is how two scientists reacted to the discovery:

> "The results have drastic consequences for the way we understand what the proton is made of," said Charles Perdrisat, one of the scientists who led the experiment.

> "[I know] of no model of the proton that would lead to such bizarre distributions of the electric field. For me, it's a real puzzle," said Ulf Meissner regarding the same phenomenon.

The fact that the antiquark is normally found in the periphery of nucleons is nowhere to be found in textbooks. Consequently, physicists have no other alternative to being surprised again and again whenever another aspect of this fact is discovered.

Only for Those Who Want to Dig Deeper

The discussion found in this chapter is based on the fact that quarks, like electrons, are Dirac particles that conform to the Pauli principle. Therefore, it is relevant to the D and E models.

Here we will try and see the effect of those pairs that exist with lower probability. In neutrons, $d\bar{d}$ pairs are less probable than $u\bar{u}$ pairs.

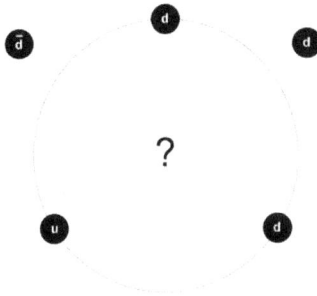

Figure 76. In a neutron with additional d quarks, the d quark is pushed to the periphery because of the Pauli principle, and a $\bar{d}$ antiquark is pushed to the periphery as indicated by Perkins's textbook.

When we have this quark combination, the Pauli principle makes the third d quark uncomfortable. It finds it "crowded" near the other two d quarks, and so we need more energy to produce a $d\bar{d}$ pair in neutrons. Therefore, this pair is less probable than $u\bar{u}$ pairs.

This "crowdedness" of d quarks stems from the Pauli principle, which pushes the other d quark to the periphery of the proton, as it is repelled by other d quarks. What follows, then, is a combination of two phenomena:

- The d quark is pushed to the periphery because of the Pauli principle.

- The $\bar{d}$ antiquark is pushed to the periphery because antiquarks are pushed to the periphery in nucleons, as indicated by Perkins's textbook.

These effects influence the distribution of charges in two opposite ways: one leads the negative charge to be pushed to the periphery, and the other leads the positive charge to be pushed to the periphery. For this reason, the two effects counteract one another, and render the effect of this combination (*ddudd̄*) on charge distribution negligible. The fact that this combination exists with lower probability makes it even more negligible.

A similar phenomenon occurs in protons.

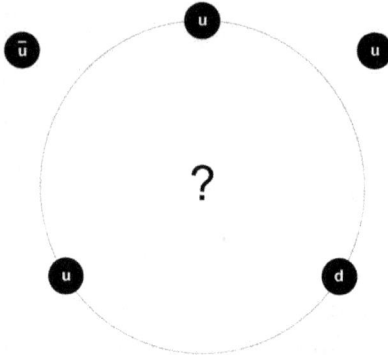

Figure 77. In a proton with additional *u* quarks, the *u* quark is pushed to the periphery because of the Pauli principle and a *ū* antiquark is pushed to the periphery as indicated by Perkins's textbook.

Exotic Particles

Dirac published his equation in 1928. According to this equation, there exists an antiparticle for each kind of Dirac particle. Indeed, a short time after that the existence of the positron, the electron's antiparticle, was confirmed. This finding corroborated Dirac's equation and paved the way for its acceptance. Had the positron not been found, it's most likely that the Dirac equation would have been discredited.

In 1961 Murray Gell-Mann and Yuval Ne'eman independently predicted the existence of the Ω^- particle, and it was indeed discovered three years later, thus corroborating the quark model. Had the particle not been discovered, the quark model would not have been consistent.

The existence of several bound particles was predicted by calculations that rely on QCD theory. In this chapter we will examine the fate of those particles.

According to QCD, quarks, antiquarks, and gluons attract one another and are able to form many kinds of bound combinations. The combinations must meet the condition that the bound particle is "white"; that is, it contains an equal amount of each of the three QCD colors.

Glueballs

According to QCD calculations made in the seventies, each gluon contains an equal amount of color and anticolor. Furthermore, a system of gluons can produce a "white" bound particle even in the absence of quarks.[159] Such a bound particle that is strictly composed of gluons has been dubbed a "glueball."

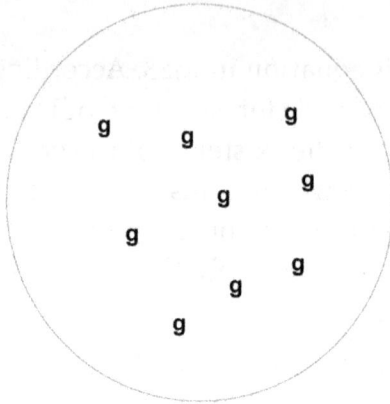

Figure 78. Glueballs made only of gluons that contain an equal amount of each color ought to exist according to QCD, but they were never discovered.

As early as in the 1970s, scientists believed that it would be possible to prove the existence of glueballs by using particle accelerators that existed at the time, but none were ever found. All that was left was a pile of excuses that explain why it was so difficult to detect their existence.

_____

159 en.wikipedia.org/wiki/Glueball. *"In particle physics, a glueball is a hypothetical composite particle. It solely consists of gluon particles, without valence quarks. Such a state is possible because gluons carry color charge and experience the strong interaction. Glueballs are extremely difficult to identify in particle accelerators, because they mix with ordinary meson states. Theoretical calculations show that glueballs should exist at energy ranges accessible with current collider technology. However, due to the aforementioned difficulty, they have (as of 2010) so far not been observed and identified with certainty."*

Dibaryons

Dibaryons, which are sometimes called hexaquarks, are hypothetical bound particles that contain six quarks, and that were predicted by QCD-based calculations.

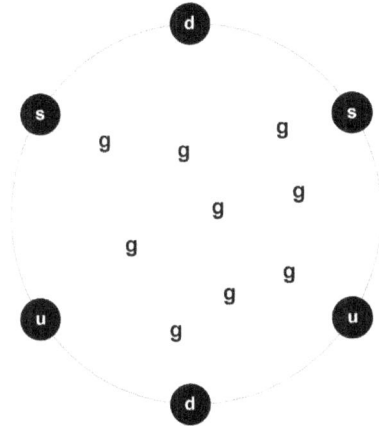

Figure 79. A *uuddss* dibaryon,
supposedly a stable particle according to QCD.

The existence of dibaryons was predicted in the seventies, and some even showed how they might be stable particles, such as *uuddss* dibaryons.[160] Despite these predictions, however, no dibaryons were ever discovered in experiments designed to detect them.

Strange Quark Matter

QCD-based calculations show that it is possible to create a stable atomic nucleus, which contains not only protons and neutrons,

---

160 R.L. Jaffe, *Perhaps a Stable Dihyperon?* Physical Review Letters **38** 195, (1977).

but also other s-quark baryons, such as the baryon $\Lambda(1116)$, which is composed of the quarks uds.[161] A single $\Lambda(1116)$ baryon is unstable, but according to QCD this baryon is stable when it is inside a large nucleus. Matter composed of such "nuclei" is known as strange quark matter, or SQM.

The fervent search for strange quark matter went beyond particle accelerators. Materials sent from various places on earth were examined in order to find out whether they contained SQM. Even lunar rocks were examined in order to find this strange matter, but none was ever found, despite the tremendous efforts invested in uncovering it.[162]

Even though strange quark matter was never found, a series of biannual conferences takes place that is devoted to the subject. Readers who missed the fourteenth convention that took place in July 2013 at Birmingham University in England are invited to follow CERN publications so as to not miss the next convention, or to occasionally Google the phrase "international conference on strangeness in quark matter."

Pentaquarks

Bound particles that consist of four quarks and an antiquark, whose bond is based on the strong force, are called "pentaquarks." These particles were predicted by QCD theory almost thirty years

---

161   E. Witten, *Cosmic separation of phases*, Phys. Rev. **D30**, 272 (1984).

162   K. Han *et al.*, *Search for Stable Strange Quark Matter in Lunar Soil*, Phys. Rev. Lett. **103**, 092302 (2009). *"No strangelets were found in the experiment."*

ago,[163] and feverish search efforts continue to this day. The bound particle of a meson and baryon, whose binding force is based on the nuclear force, is a residual force that is much weaker than the strong force, and this particle is not considered a pentaquark.

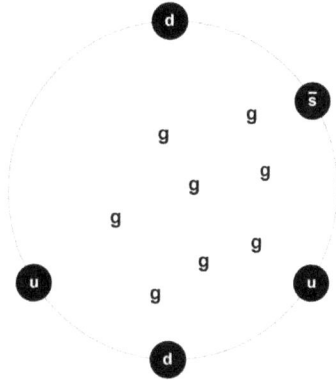

Figure 80. Pentaquarks made up of four quarks and an antiquark that are strongly bound should exist according to QCD, but they were never discovered.

Hypothetical pentaquarks were actually classified in the past, and scientists published different predictions about their properties. In 1997 it was revealed that a pentaquark composed of the combination *uudds̄* ought to be relatively stable and should have a mass of 1530 MeV. This pentaquark was referred to as θ^+.

In fact, the θ^+ particle is not an authentic pentaquark, because it is an unbound state whose mass is too high, and thus it fails to comply with condition that a pentaquark be a state bound by the strong force. But in any event, it's interesting to see how the story continues.

---

163 M. Praszałowicz (1987). M. Jeżabek, M. Praszałowicz, ed. Proceedings of the Workshop on Skyrmions and Anomalies, Krakow, Poland, 1987. World Scientific. p. 112.

Suddenly, in 2003, scientists working at the LEPS particle accelerator in Japan announced that they found a θ⁺ with a great degree of certainty. Immediately after this publication other groups of scientists working with other particle accelerators announced that they had found additional pentaquarks.

Strangely, however, each and every one of these publications turned out to be unfounded. According to a summary circulated by the PDG, the agency authorized to confirm the existence of new particles:

> … There are two or three recent experiments that find weak evidence for signals near the nominal masses, but there is simply no point in tabulating them in view of the overwhelming evidence that the claimed pentaquarks do not exist… The whole story—the discoveries themselves, the tidal wave of papers by theorists and phenomenologists that followed, and the eventual "undiscovery"—is a curious episode in the history of science.[164]

Well-established theories have acquired their prestige by fulfilling predictions on the existence of new particles. Logically speaking, then, the reverse argument should also be true: the persistent non-discovery of particles that ought to have been discovered by a theory implies that there are holes in the theory and its validity is undermined.

---

164 C.G. Wohl (LBNL), *Pentaquarks*, 2008.

Unconfined Mesons

The logic according to which QCD predicts the existence of pentaquarks begets yet another contradiction to QCD. According to QCD, quarks attract each other and the force that operates between them grows stronger as they draw farther away from each other. Therefore, when a quark seeks to leave a baryon, the force that binds it becomes ever-greater the more it draws away from its fellow quarks, and in fact becomes virtually infinite. This explains, according to QCD, why quarks remain confined inside a baryon.

However, mesons made up of a quark and an antiquark can leave a baryon with relative ease. And indeed when an energetic photon hits a baryon, the baryon emits a meson with little effort.

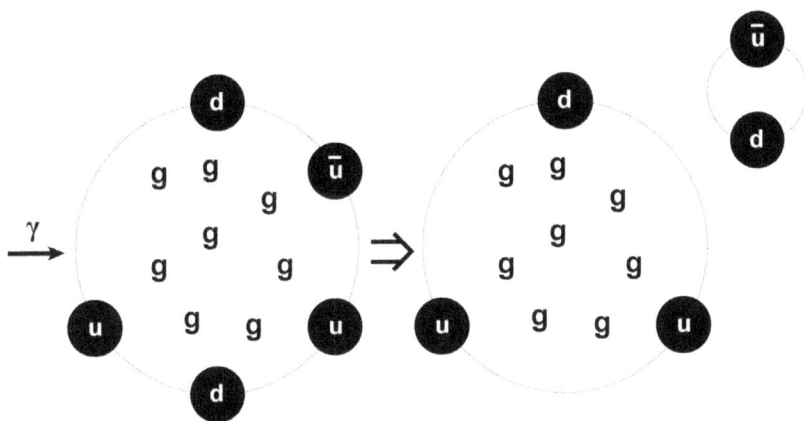

Figure 81. A photon hits a neutron and isolated proton and π^- are produced. According to QCD, quarks and antiquarks all attract each other. A single quark is confined in the proton, and an antiquark as well, although quark-antiquark pairs are able to leave the proton with relative ease.

Why does this happen? Antiquarks are also attracted to quarks according to QCD. How is it, then, that a meson, which is composed of a quark-antiquark pair, can actually leave the baryon? Why doesn't it remain confined?

QCD allegedly has an answer to that question, but it seems to contradict several basic laws of physics. According to QCD, the colors of the quark and antiquark cancel each other out, and, as a result, the force that operates on them is eliminated, and for this reason the quark and antiquark are able to leave the baryon in the form of a meson.

Does this explanation make sense? What it implies is that according to QCD the force that operates inside the baryon on a single quark is huge and is in fact practically infinite—furthermore, the force operating on a single antiquark is also practically infinite, but somehow the force operating on a quark-antiquark pair is canceled out. This explanation stands in stark contradiction to a basic physical law that says that the force operating on two objects is the sum of the force operating on one object and the force operating on the other.

By the way, it is interesting to point out that a meson is made of a spatially separated quark-antiquark pair. Therefore, the aforementioned QCD "explanation" violates yet another law of physics called "locality."

The mathematical framework of QCD is indeed very elaborate compared to, say, electromagnetic theory. However, it remains entirely unclear whether QCD can consistently explain this effect. In fact, since according to QCD pentaquarks should exist, and they are a strongly bound baryon and meson, then it implies that according to QCD mesons should be confined inside the proton. But this is not happening. Therefore, the easy exit of a meson from a baryon is yet another contradiction to QCD.

As for the other models, the D and E models can explain why quarks are confined and mesons aren't. Since quarks repel one another due to their having the same strong (negative) charge, and since the antiquark is attracted to them but is repelled by the baryon core due to its having a strong (positive) charge, the quark and antiquark are only attracted to the baryon by a residual force and are therefore able to leave as a pair through the investment of a relatively small amount of energy.

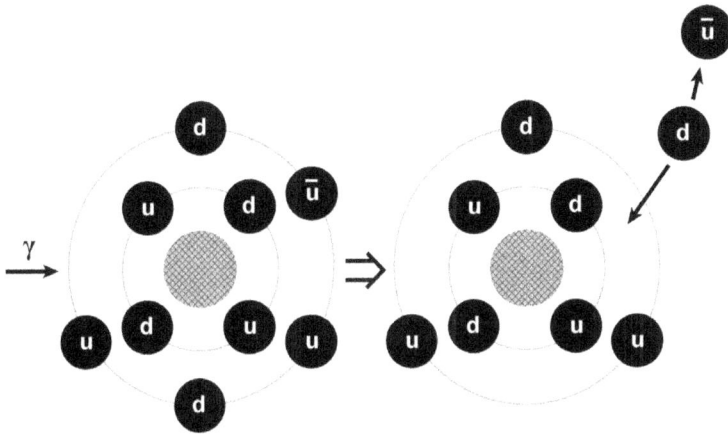

Figure 82. According to version E of the liquid drop model, quarks repel each other and are attracted to the center, and the antiquark is repelled by the center and is attracted to the quarks. Therefore, the quark and antiquark are only bound to the proton by a residual force.

A Summary of
Unexplained Phenomena Thus Far

Table 8. Problematic phenomena
associated with the repulsion forces between quarks.

Contradicting Phenomena	Suspicious Phenomena
17. The antiquark is situated in the outermost region of the nucleon.	
18. The negative charge inside the neutron tends to be found in the outermost region of the neutron.	
19. The positive charge inside the proton tends to concentrate in the outermost regions of the proton.	
20. No glueballs were ever found.	
21. No dibaryons were ever found.	
22. No strange quark matter was ever found.	
23. No pentaquarks were ever found.	
24. Mesons are not confined to nucleons.	

So Why Do Scientists Still Believe in QCD?

A List of Explanations

As part of my research for this book, I asked several scientists why they subscribe to QCD. I received several answers, some of which were long, detailed, and profound.

Most of the arguments presented were discussed in previous chapters, to wit:

- There is no other explanation for the existence of Δ^{++} and Ω^- particles.

- The Landau pole.

- The three jets event.

- Gerardus 't Hooft (awarded a Nobel Prize in 1999 for his work on electroweak forces) replied by saying that there exists no theory other than QCD that can be consistent with special relativity.[165]

Other arguments were also given, such as quark confinement, and the fact that we never find hadrons with two or four quarks. As far as these arguments are concerned, scientists admitted the validity of the alternative explanation found in the D and E models.

Three topics are yet to be reviewed in this book and have been raised by several scientists. These topics will be addressed in this chapter.

---

165   Private letter. August 4, 2011.

Lattice QCD

QCD equations are so complex that it is simply impossible to employ them directly. In 1974 Kent Wilson developed an approximate calculation method called lattice QCD.[166] According to this method, it's possible to select a "lattice" of points. The denser the lattice, the more accurate the calculation and the more it agrees with the original QCD equations, although that also leads to a slower calculation process.

The lattice method has been known for quite some time, and it is used to find a computational solution for complex differential equations. It began to be commonly used with the advent of electronic computers.

When scientists employ lattice QCD techniques, they enjoy a tremendous amount of freedom in selecting physical parameters, such as particle mass and the type of lattice with which they work. In this fashion, they can retroactively "explain" the masses of measured particles by selecting their parameters and lattices by means of trial and error. Nevertheless, it was only in 2008 that they finally managed to explain the mass of the proton by using lattice QCD, with an accuracy of a few percent.[167]

Should that be regarded as a triumph? Why, in the decades that elapsed since the invention of this method and until the article was published in 2008, have scientists engaged in dozens of computational experiments and never gotten it? If they had gotten it, we would have announced it sooner. Thirty-four years had to pass

---

166 K. Wilson, *Confinement of quarks*, Phys. Rev. D **10** (1974).

167 S. Dürr, Z. Fodor, J. Frison *et al.* *Ab Initio Determination of Light Hadron Masses*, Science vol **322** (2008). p. 1224–1227.

before thousands of scientists around the world finally managed to find the right combination of parameters to obtain the slightly inaccurate mass of protons and several other light hadrons.

Freeman Dyson mentioned one occasion in the fifties when he presented Enrico Fermi with a model that demonstrates how, by selecting certain parameters, one can explain a certain graph associated with hadrons. Fermi rejected the idea and quoted von Neumann: "With four parameters I can fit an elephant, and with five I can make him wiggle his trunk!" In this way Fermi expressed his dislike for multi-parameter models. Such models can explain everything that has already been measured, but they turn out to be useless when one tries to employ them to predict new phenomena.

When I asked several physicists why they believe in QCD, some of them mentioned the "successful" explanation for light hadron masses that was achieved in 2008 as proof that QCD works well and is commensurate with scientific data. But when they were asked whether lattice QCD has any *prediction* that was later confirmed by experiment, they admitted that not a single one exists ("nothing to write home about," as one subtly put it).

The Prediction of Hadron Masses

One of the successes of QCD scientists was a rather good prediction of yet unmeasured hadron masses. They produced that prediction by examining other particles that have similar properties and a known mass, and then formulated an educated guess of the new particle's mass.

But a method that employs a similar principle had already been developed in the sixties by Gell-Mann[168] and Okubo[169] called the Gell-Mann-Okubo mass formula.[170] This formula precisely corresponds to measured masses with an accuracy level that is similar to that achieved by QCD, although it was developed a decade before QCD even existed, and, of course, it does not rely on QCD at all. It was even further refined on a later occasion.[171]

π^0 Decay

The pion π^0 is made up of a combination of $u\bar{u}$ and/or $d\bar{d}$ quarks. It is the lightest form of meson, and it has three principal states.

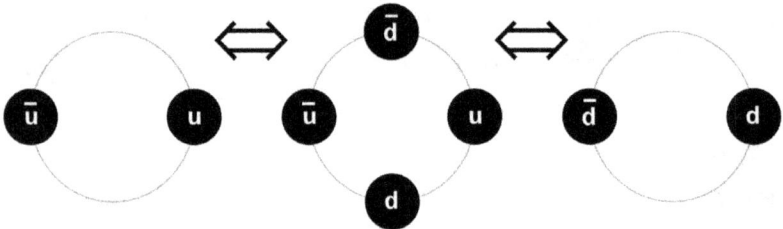

Figure 83. The three most common states of the π^0 pion.
The pion moves from state to state every fraction of a second.

---

168 M. Gell-Mann, *The Eightfold Way: A Theory of Strong Interaction Symmetry*, California Institute of Technology Synchrotron Laboratory Report CTSL-20 (1961).

169 S. Okubo, *Note on Unitary Symmetry in Strong Interactions*, Prog. Theor. Phys. 27 (1962).

170 en.wikipedia.org/wiki/Gell-Mann–Okubo_mass_formula.

171 E.g.: *Gell-Mann Okubo Mass Formula Revisited*, L. Burakovsky and T. Goldman, 1997. arxiv.org/abs/hep-ph/9708498.

The pion moves from state to state at tremendous speed. It has other, rarer states as well, which include other quark-antiquark pairs.

It was already known in the 1940s that this pion decays into two photons. As photons are normally formed only by electromagnetic interactions, scientists believed that the pion decays because of the electric charge of its quark and antiquark.

Quantum field theory, which was primarily developed in the 1940s and 1950s, provides tools that allow us to calculate electromagnetic processes in general, and the time of the particle and antiparticle's decay in particular. For example, it makes it possible to identify the nature and duration of the decay observed in a particle composed of an electron-positron bound state, known as a positronium. Indeed, quantum field theory, which assumes that the decay of a particle and its antiparticle stems from electric interaction, provides a wonderful estimate of the positronium's lifetime.

When field theory equations were applied to the π^0 pion, it turned out that it decayed at a rate approximately nine times faster than what had been predicted by the equations. And then, when QCD was invented, the "three colors" were introduced into the equation and were used to explain this difference.

I was presented with this argument by two scientists who used it to try to convince me of QCD's validity, and it is also mentioned in academic literature as supporting evidence for QCD.[172] According

---

172 Michael E. Peskin and Daniel V. Schroeder, *An Introduction to Quantum Field Theory* (Perseus, Reading, MA, 1995), p. 676.

to Fritzsch, this calculation was an important factor in the development of the theory.[173]

Are readers convinced? Well, let's take a look at the calculations that appear in Peskin and Schroeder's book.

It turns out that the calculations they had made in order to identify the duration of the pion's decay, which uses field theory equations, did not assume that the pion exists in many cases where it has more than one quark-antiquark pair. When the pion exists in a state that includes more than one pair, such as the one appearing in the middle of Figure 83, then when "decay" occurs and a quark is annihilated with the antiquark, the pion still remains because another quark-antiquark pair (or more than one pair) exists.

When this factor is taken into consideration, the pion's projected duration of decay should be substantially longer, and the difference between the time calculated using field theory methods and the actual result ought to be much greater than 1:9. Therefore, the three colors fail to explain this phenomenon.

It follows, then, that the explanation for the pion's decay duration is not exactly a resounding success of QCD, but rather it rests on shaky foundations.

What better alternative explanation can we have for the fact that the pion decays much faster than expected from the electric interaction between the quark and antiquark? Remember that photons interact with strong charges and not just with electric charges? We

---

173 Harald Fritzsch, *The history of QCD*, CERN courier, September 27, 2012. *"In 1971–1972, Gell-Mann and I worked at CERN. Together with William Bardeen we investigated the electromagnetic decay of the neutral pion into two photons. It was known that in the quark model the decay rate is about a factor nine less than the measured decay rate—another problem for the quark model."*

considered this forgotten phenomenon in the chapter titled "The Photon and the Strong Force." It's possible, then, that a correct theory that explains the interaction of photons and strong forces would also be able to explain how the π^0 particle decays into a pair of photons by means of the strong rather than electric force.

QCD and the Chronological Test

In the beginning of this book I presented the chronological evidence test designed to test the reliability of a theory. According to this test, we need to honestly ask ourselves whether the theory to which we currently subscribe would have been invented had the experimental data known today been available before the birth of that theory.

When QCD was invented, scientists thought that protons and neutrons were made up of only three quarks. They were convinced that three quarks inside light baryons ought to be situated in the innermost shell; therefore they must lack orbital spin, and that's where we found a contradiction to the Pauli principle, which forms one of the foundations of physics.

Today we know that a large portion of the nucleon's mass is not carried by external quarks. We also know that in experiments that measured high-energy proton-proton collisions, we observe a great number of elastic collisions, which indicate the existence of a massive rigid core inside the proton. In addition, we know that the cross section graph of proton-proton collisions rises at very high energy levels, a phenomenon that may be explained by inner quark layers, similarly to an atom with more than one shell. Moreover, since then we have encountered the proton spin crisis and other findings indicating that the quarks inside nucleons carry a significant orbital angular momentum.

And now, let's assume that all of these data were already available in the sixties, before QCD was born. Would scientists honestly have invented this theory, which is unlike anything else known in physics, and which contains an endless amount of fantastic assumptions? Not only do these data resolve the supposed contradiction discovered in the sixties, they even manage to completely contradict QCD itself!

Several years ago I gave a short article on the subject to a senior particle physicist at the Weizmann Institute. The reply I got was that "had this explanation been presented forty years ago, it would have been relevant at the time, but it's too late now." This senior physicist preferred to stay anonymous, and his reply might attest to how eager scientists are to uncover the scientific truth.

THE ALTERNATIVE HISTORY OF PARTICLE PHYSICS

Magnetic Monopoles

The main subject matter of the book is not a dispute among scientists, but rather a community of physicists that has been ignoring dozens of repeated failures of the theory to which it subscribes. This theory glorifies its "achievements" in ways that do not allow free academic debate. It obtains gigantic, billion-dollar budgets every year at taxpayers' expense.

I do hope that by now readers are convinced that a key part of the Standard Model is unreasonable and rests on shaky ground, or at the very least is highly suspect. If readers aren't convinced, I apologize. But if upon reading this book anyone has become suspicious of QCD, I would like to tell a story.

Over the last century, scientists have been engaged in a corresponding line of research in an attempt to explain the structure of the proton in a way that contradicts QCD. This group of scientists includes two Nobel Prize laureates, Paul Dirac and Julian Schwinger, and many other scientists, including Tetsuo Sawada, the vice president of Atoms for Peace, and Eliyahu Comay, the scientific advisor for this book, who is responsible for making sure that claims relating to physics mentioned in this book are accurate. The ideas used by this alternative scientist group continue to attract scientists to this day.[174]

174　Luis N. Epele, Huner Fanchiotti, Carlos A. García Canal, Vasiliki A. Mitsou, Vicente Vento, *Looking for magnetic monopoles at LHC with diphoton events*, European Physical Journal Plus, **127** 2012.

The story begins in 1931 with an article published by Dirac on magnetic monopoles. But before we hear the actual story, let's first define two basic terms in electromagnetism.

Electric and Magnetic Dipoles

An axial magnetic dipole (or "magnetic dipole" for short) is formed when an electric current moves along a circle or when an electrically charged elementary particle possesses spin. Two magnetic poles exert a force on each other. The direction of the force depends on the precise arrangement of the dipoles, and it is usually not aligned on the straight line that connects them. This force is known as "tensor force." I am assuming that anyone who has ever played with two magnets has sensed this force with his or her own hands.

An electric dipole is formed when two opposite electric charges are situated next to one another, and the sum of their charges becomes zero. When the charges are measured from a distance, the separate electric charges are barely perceptible, and any two such dipoles act on each other very similarly to two magnetic dipoles.

In nature we have point electric charges, but we never find a magnetic charge called a magnetic monopole (or simply "monopole"). The basic equations of the theory of electricity, Maxwell's equations, are based on the assumption that a magnetic monopole does not exist.

Dirac's Monopoles

In 1931 Dirac wrote an article[175] that describes the properties of a particle carrying a magnetic monopole. Dirac's article sought to describe this monopole because he wanted to explain why electric charges are characteristically discrete (or "quantized"). And Dirac indeed showed that if there had been a monopole somewhere in the universe, then the electric charge found in nature would have to be in discrete rather than continuous form. Today it is universally accepted that electric charges are indeed quantized, but, despite an abundance of experimental efforts that persist to this day, not a single monopole has ever really been discovered in the laboratory.

Let's examine the structure of Dirac's article, without going into the equations themselves. The way in which Dirac devised the theory of monopoles may shed light on the historical development of particle physics all the way to the present. That is why we should try and understand the concepts of this important article.

Dirac began by describing the magnetic charge through the use of several equations that were similar to Maxwell's, while using a magnetic rather than an electric charge, which underwent a certain transformation to ensure the equations' consistency. So far so good. The equations remained consistent, and they described the properties of the new charge, which behaved in a very similar way to an electric charge; this outcome was achieved without any particular problem. It is important to note that Maxwell's equations are written in terms of electric and magnetic fields.

---

175   P. A. M. Dirac, *Quantised Singularities in the Electromagnetic Field*, Proc. Roy. Soc. (London) **A 133**, 60 (1931).

We will return to this transformation later on when we discuss how it relates to the deuteron's tensor force.

Dirac then proceeded to examine quantum mechanical systems. Here it is mandatory to use a quantity called 4-potential. (The electric and the magnetic fields are derived from this 4-potential.) Dirac introduced the 4-potential and examined what would happen to a particle with a magnetic monopole when situated near an electric charge. Dirac made another tacit assumption where the same 4-potential is used for the charge and the monopole.

And that is where the equations collapsed—because of all the constraints imposed on them by Dirac. Dirac understood that the equations he created had become unphysical under these conditions. And in order to make these equations useful, Dirac invented the concept of "Dirac strings," which are connected to every monopole. According to this concept, the points in space where the equations fail are situated on a "string." Moreover, no electric charge may be located on such a string.

Dirac's strings are lines that spread out in space and the imposed nonexistence of electric charges on them is known as "Dirac's veto."

The main result of his article was that electric charges are "quantized." Another important result was that the ratio between electric and magnetic charges is 2:137—namely, the elementary magnetic charge is much stronger than the elementary electric charge. This numerical ratio represented very strong forces that are unparalleled among known forces. Note that the interaction between a pair of electric charges is *quadratic* in the charge. Hence, the corresponding monopole interaction is nearly five thousand times stronger than a basic electric interaction.

Dirac, along with other scientists, was unhappy with the string concept, and there were quite a few attempts at formulating a consistent monopole theory that is devoid of Dirac's strings.[176,177,178]

Monopoles and Protons

Dirac continued in his pursuit of the monopole concept for many years. In the meantime, other scientists were busy uncovering the structure of the proton. In 1933, measurements indicated that the proton was in fact not a Dirac particle and was therefore not elementary. In 1934 Pauli and Weisskopf reconstructed the Klein-Gordon equation to which Dirac had objected his whole life. Scientists believed that Dirac's objections were political rather than professional, and they stemmed from the fact that these equations show a massive elementary particle that is not a Dirac particle. In 1935 Yukawa formulated his theory, which assumed that the proton and the neutron were elementary particles and another elementary particle existed—the Yukawa particle—that conformed to the Klein-Gordon equations and carried the nuclear force that keeps neutrons and protons glued together inside the atomic nucleus. When a particle with properties similar to those predicted by Yukawa was discovered in 1947, the scientific community accepted Yukawa's theory and the Klein-Gordon equation was back at the forefront once more.

---

176 Richard A. Brandt and Joel R. Primack, *Avoiding "Dirac's veto" in monopole theory*, Physical Review **D15** (1977). p. 1798–1802.

177 K Hirata, *Classical lagrangian theory of Dirac's monopole: Avoiding Dirac's veto*, Physics Letters **B81**, 1979. p. 169.

178 Tai Tsun Wu, Chen Ning Yang, *Dirac's monopole without strings: classical Lagrangian theory*, Physical Review **D14**, 1976. p. 437–445.

I believe that at that point Dirac felt physics was heading in the wrong direction, and in 1948 he published another article on magnetic monopoles.

Dirac's second monopole article is a lot more elaborate than its predecessor, and it surprisingly ends with the announcement that magnetic monopoles may be found inside protons.[179] By so doing Dirac confronted the mainstream opinion of scientists at the time, who usually favored Yukawa's theory.

Yukawa was awarded a Nobel Prize in 1949 for his theory, and it was not much longer than a decade later that Yukawa's theory was found to be incorrect. It was finally made clear that the proton is not an elementary particle, the neutron is not an elementary particle, and even the particle that was thought to be the Yukawa particle wasn't elementary either. It's also an accepted fact today that there is no elementary particle that satisfies the Klein-Gordon equation.

When it was discovered that the proton and neutron were made up of quarks, Julian Schwinger tried to implement the concept of monopoles as components of protons.[180] He formulated a theory whereby quarks conformed to equations that were very similar to Dirac's monopole equations. But one of the outcomes of Dirac's assumptions was that if quarks carry a magnetic charge, then they should have produced a relatively strong electric dipole moment. However, scientists who examined an electron traveling near a neutron were unable to detect this dipole, and this was probably the reason Schwinger's idea also failed.

---

179 P.A.M. Dirac, *The Theory of Magnetic Poles*, Phys. Rev. **74**, 817 (1948). *"...We might suppose that elementary particles with poles form an important constituent of protons..."*

180 Julian Schwinger, *A magnetic model of matter*, Science Vol **165** (1969).

What Does Gell-Mann Want?

In the 1960s, Eliyahu Comay, the scientific advisor for this book, earned his master's degree from the Hebrew University in Jerusalem. As a graduate student he worked on applying the Wigner-Racah calculations to nucleonic configurations in the shells of atomic nuclei. In this fashion he sought to explain the masses of light atomic nuclei.

In the course of his studies Comay encountered the staggering similarity between the graph representing inter-nucleonic forces and the graph that depicts the forces that operate between noble atoms. He decided to regard this similarity as a strong indication that the underlying forces of these systems were similar as well. This idea required that the internal structure of protons and neutrons was similar to that of the atom, with quarks assuming the role of electrons, and with strong forces replacing electromagnetic forces. As stated earlier, in this book, this idea is summarized as the "liquid drop model."

Comay was further encouraged to pursue this line of reasoning because of the many similarities that existed between strong and electromagnetic forces. For example, the forces operating between quarks were parity-conserving, which was also true for electromagnetic forces. At the time, before the birth of QCD, there were other scientists who thought the electric force was similar to the strong force.[181,182,183]

The A1 Meson

In 1970, Comay transferred to Tel Aviv University, where he presented his doctoral advisor with the liquid drop model concept, and the analogy between electromagnetic and strong forces. Comay proposed to calculate the states of seven mesons by using mathematical formulas that were already known in atomic and nuclear spectroscopy. At the time the Israeli community of physicists was suffering from the absence of two great scientists, Yoel Racah and Amos De Shalit, who were considered to be experts familiar with such calculations and who had even developed some of their own. Racah died suddenly in 1965, and De Shalit died suddenly as well in 1969.

At that point Comay's advisor informed him that if Comay were to pursue his proposed subject, he would have no interest in being his advisor, but he was willing to have Comay work on the subject on his own. If Comay's findings were interesting enough, the advisor might consider joining his efforts.

Sure of his abilities, Comay rose to the challenge. At first he decided to show that the liquid drop model could explain the mass-

181   P. Yock, *Unified field theory of quarks and electrons*, International Journal of Theoretical Physics, Vol 2, Issue 3, p. 247–254 (1969).

182   A. O. Barut, *Proton Form Factor, Magnetic Charges, and Dyonium*, Phys. Rev. **D3**, 1747 (1971).

183   T. Sawada, Phys. Lett. **43B**, 517 (1973).

es of several mesons that had been measured by that time. He selected a group of seven mesons, all of which were composed of u and d quarks, and tried to see whether their masses corresponded with his ideas.

Comay used a computer, which was not a popular tool at the time. In 1970 one needed a spacious hall to contain a computer and the equipment that supplements it. Another "quirk" of the Jerusalemite, his Tel Avivi advisor must have thought.

After punching thousands of cards while using the "modern" Fortran programming language, the results were finally in. They confirmed with good accuracy the mass of six out of the seven mesons selected by Comay. The seventh meson, however, known as the A1 meson, had yielded results that were completely off the mark.

Comay persisted in his struggle to find out why this wayward meson failed to produce good results, but to no avail.

Today we know that scientists in the seventies had made a huge mistake when they measured the A1 meson (see Figure 84). In fact, Comay's calculations did not show that he was wrong, but instead they could have formed the basis for predicting the correct mass of this meson!

However, being as inexperienced as he was at the time, Comay gave up, went back to the advisor, and took it upon himself to find another topic for his doctoral thesis. And so, in 1972, he found himself sitting in front of his advisor and waiting to hear what his new doctoral thesis would focus on. The advisor, who was more than a bit pleased with his "success," then declared, "Now we're going to do what Gell-Mann wants us to do!"

138   Reviews of Modern Physics · January 1970

Data in parentheses have not

3 ETA (1060) BRANCHING RATIOS

```
R1        ETA (1060) INTO (P1 PI)/IK KBAR)
R1        (2.5)  OR LESS        CRENNELL   66 HBC    90 PCT CONF LEV   7/66
R1        1.0   .0.6      0.3   LAI        66 HBC    6 PI= P            11/68
```

REFERENCES FOR ETA(1060)

```
RIGI        62 CERN CONF 247    A RIGI,S BRANDY, R CAPRARA +         (CERN)
BINGHAM     62 CERN CONF 240    H H BINGHAM,H BLOCH +  (PARIS+EC POLY+CERN)
ERWIN       62 PRL 9 34         ERWIN,HOYER,MARCH,WALKER,WANGLER   (WIS+BNL)

BALTAY      64 DUBNA CONF 1 409 BALTAY,LACH,CRENNELL,OREN,STUMP +(YALE+BNL)
BARMIN      64 DUBNA CONF 1 433 BARMIN,DOLGOLENKO,YEROFEEV,KRESTNI+ (ITEP)

CRENNELL    66 PRL 16 1025      CRENNELL,KALBFLEISCH,LAI,SCARR,SCHU+ (BNL)
HESS        66 PRL 17 1109      ADAM,HARDY+KIRZ+MILLER                (LRL)
  HESS REPLACES  PRL 9 460      ALEXANDER,DAHL,JACOBS,KALNFLEISCH +  (LRL)

BARLOW      67 NC 50A 701       +LILLESTOL+MONTANET+(CERN+CDF+IP+LIVERPOOL)
DEUSCH      67 PL 29 B 357      +FISCHER,GORBI,ASTBURY,+MICHELINI+(ETH+CERN)
DAHL        67 PR 163 1377      +HARDY+HESS+KIRZ+MILLER              (LRL)

ALITTI      68 PRL 21 1705      +BARNES,CRENNELL,FLAMINIO,GOLDBERG,+ (BNL)
HUANG       69 NC 61 A 325      T.F.HOANG                           (ANL)
LAI         68 PHILAD.CONF.P.303 KWAN WU LAI                        (ANL)
PHELAN      68 THESIS           JAMES J. PHELAN    (ANL+ST.LOUIS UNIV)
  ALSO 68 PPL 21 316            HOANG,EARELY,PHELAN,ROBERTS+(ANL+CHIC+NDAM)

AGUILAR-    69 PL 29 B 241      M.AGUILAR-BENITEZ,J.BARLOW,+        (CERN+CDF)
  ALSO BARLOW 67
```

A1(1070)   10 A1 MESON (1070, JPG=1+-) I=1

Citation: K. Nakamura *et al.* (Particle Data Group), JPG **37**, 075021 (2010)

$a_1(1260)$ $I^G(J^{PC}) = 1^-$

See also our review under the $a_1(1260)$ in PDG Physics, G **33** 1 (2006).

$a_1(1260)$ MASS

Figure 84. The mass of the A1 meson was estimated at 1070 in 1970 (top), but as early as the 1980s it was discovered that it was actually equal to 1260 (bottom).

Comay had and still has the greatest respect for Murray Gell-Mann, the 1969 Noble Prize laureate. But he didn't think that scientists were supposed to seek scientific truth inside the mind of any particular scientist, even the mind of Gell-Mann. Comay rejected the offer and later completed his thesis in the nuclear physics department.

Monopoles, Once More

In 1982 Comay was at the University of Michigan when he found an article that reported the discovery of a magnetic monopole. The discovery turned out to be completely unfounded, but it did pique Comay's interest. He knew from his university days that Maxwell's equations, which are highly efficient equations, are based on the assumption that monopoles did not exist. Nevertheless, he made an effort to try and understand the theoretical aspects of monopoles.

To his surprise, it was very hard to find any convincing theoretical material on the subject, and he was forced to go back to Dirac's original 1931 article. While reading the article, Comay noticed that Dirac used an unnecessary assumption, which implies that the magnetic monopole interacts with the fields generated by the electric charge. He marked a small x near the line where the assumption appeared and finished reading the article.

The usage of strings made by Dirac to unravel the mathematical Gordian knot seemed to Comay to be artificial and wrong. The theory of electrodynamics had no need for strings! He decided to try and develop a new set of monopole equations that lacked this superfluous assumption. He instead used the variational principle, a principle commonly accepted in many fields of physics in general and particularly in electrodynamics.

Several weeks later the new monopole equations were complete, this time without the flaws found in Dirac's monopole equations. According to Comay's equations, there was no connection between the size of the electric charge and that of the magnetic charge. He noticed that, in fact, the new monopole equations presented two new types of fields, one analogous to the electric field and the other to the magnetic field. Between the two new fields and the electric and magnetic fields one found no direct interaction whatsoever, but they did share something in common: the photon (and its fields) interacted with monopoles as well, and not just with electric charges.

At this point Comay recalled the rather strong interaction of photons with protons and neutrons, and considered Sakurai's VMD theory, which always seemed to him unlikely because it contradicted special relativity. Quarks, then, had conformed to his new equations, and the strong charge was actually the same one described by the new monopole equations. The photons interacted with the strong charge to produce the hadronic properties of the photon—an attribute that was a well-established phenomenon.

Comay, who specialized in nuclear physics that concerned internucleonic forces, recalled the unsolved problem of the deuteron's tensor force. Here I remind readers again that electric charges cannot explain this force, among other reasons because in nature it appears with an opposite sign to that derived from electromagnetic field equations. Comay examined whether the tensor force stemmed from the fact that quarks carried a strong force that conformed to his equations. When he tested the new equations and developed his "strong dipole" idea, he found out that the tensor force actually appeared with the right sign, owing to the transformation Dirac had introduced into the original equations. It turned out that the new equations were compatible with that phenomenon as well!

Sounds great so far, right?

Astonished by his discovery, he turned to a colleague at the University of Michigan and offered him the opportunity to partner up. The colleague refused and suggested that Comay hold a seminar dedicated to the subject. The seminar was attended by local experts on monopoles who turned out to be skeptical and rejected the new idea.

Comay's article, which describes the mathematical structure of monopole properties, was published in 1984, and did not address their conformity to the monopole carried by quarks.[184] In this fashion, he was able to describe the corrections he had found for Dirac's equations without mentioning that they contradicted QCD.

Later Comay showed that it was possible to obtain the same monopole equations without using the variational principle, but rather on the basis of a different assumption.[185]

Comay wasn't the only one who recognized the monopole idea and the liquid drop model as an explanation for the similarity between van der Waals forces and the nuclear force.[186] But even Tetsuo Sawada, an esteemed nuclear physicist and one of the senior figures of the International Institute for Sustainable Peace, was unable to publish his reservations about QCD in a leading scientific journal, and he had to make do with a minor publication that was intended, most likely, to be read at some point in the distant future. In his article Sawada showed that the binding energy between nucleons was similar to the van der Waals force, rather than the force derived from QCD.

---

184 E. Comay, *Axiomatic deduction of equations of motion in classical electrodynamics*, Il Nuovo Cimento, vol **80B**, 159 (1984).

185 E. Comay, *Charges, monopoles and duality relations*, Il Nuovo Cimento B, vol. **110**, 1347 (1995).

186 Tetsuo Sawada, *Is the nuclear force short range?* arXiv:nucl-th/0307023, (2003). *Therefore the value of the strength C of the Van der Waals potential supports the magnetic monopole model of hadron rather than the QCD.*

The Loneliness of
Searching for Scientific Truth

Since an established and leading group of scientific journals did not accept any scientific articles that contradicted the Standard Model, Comay published his entire theory through other scientific journals, usually in small portions, with each article examining a certain aspect of his theory. In 2012 a long review article was published that summarized the main points of the theory.[187] In a world that contains a functioning scientific community, scientists would have taken the ideas published in those articles seriously, and would have tried to uncover the flaws found in arguments or mathematics that appear in them. In reality, however, even though many people were exposed to Comay's articles (thousands downloaded them), not a single attempt of this sort made its way to mainstream scientific journals.

The next section is rather gossipy in nature and it does not directly relate to the main subject matter of this book. Readers who aren't interested in such things may choose to skip the next part, but if anyone is interested, here are a few unusual episodes that may shed more light on Comay's character and on that of academia in general.

---

187 E. Comay, *The Regular Charge-Monopole Theory and Strong Interactions*, Electronic Journal of Theoretical Physics. Vol **26**, 93 (2012).

The Aharonov-Bohm Effect

In 1959 Yakir Aharonov and David Bohm published an interesting article that described the Aharonov-Bohm Effect.[188] The article was considered by many to be Aharonov's greatest achievement, especially because it made extensive use of a branch of mathematics known as topology.

Using this new tool they had developed, Aharonov and Bohm predicted the existence of two effects, known as the magnetic Aharonov-Bohm effect and the electric Aharonov-Bohm effect. The magnetic effect was discovered in an experiment several years later, and the pair's work gained a significant deal of interest among physicists.

In 1987 Comay published an article that argued that the electric Aharonov-Bohm effect could not exist, as it contradicted the law of conservation of energy.[189] He later published another article that argued that the magnetic effect could be explained without using topology.[190]

After a while, the Tel Aviv University School of Physics arranged two lectures, one to be given by Comay and the other by Aharonov. The hall was packed, and the two went on and presented their arguments. When one of the physicists in the auditorium was asked who had won, he said that the discussion was fascinating, even though he wasn't able to fully understand what it was about.

---

188   Y. Aharonov and D. Bohm, *Significance of electromagnetic potentials in quantum theory*, Physical Review **115**, 485–491 (1959).

189   E. Comay, *Further comments on the original derivation of the electric Aharonov-Bohm effect*, Physics Letters **A120** 196, (1987).

190   E. Comay, *Interrelations between the neutron's magnetic interactions and the magnetic Aharonov-Bohm effect*, Physical Review A **62**, 042102, 2000.

After the "battle," Aharonov asked Comay, somewhat belatedly, for a printed copy of his article showing that there were problems with the Aharonov-Bohm idea. Aharonov has yet to provide his reply to that article.

To this day, as Comay had predicted, the electric Aharonov-Bohm effect has never been confirmed.[191,192,193]

The Hidden Momentum

In 1967 Shockley and James published an article that contained a "thought experiment" in which momentum is produced out of nothing. According to the laws of physics, momentum is a conserved quantity, just like energy. For this reason, this thought experiment had been dubbed the hidden momentum paradox.[194] William Shockley is a Nobel Prize laureate, and the paradox re-

191   E.g.: *"...there isn't any direct experimental observation of the electric AB effect..."* A.V. Ghazaryan, K. Moulopoulos, A. P. Djotyan and A. A. Kirakosyan, *Investigation of the electric Aharonov-Bohm effect in a quantum ring,* 50 years of the Aharonov-Bohm effect, Tel-Aviv University, 2009.

192   *"The existence of electric Aharonov-Bohm effect, that has not been confirmed experimentally, is a very controversial issue."* The *Electric Aharonov-Bohm Effect,* Ricardo Weder, 2010. arxiv.org/abs/1006.1385.

193   Batelaan, A.; Tonomura, A. (Sept. 2009). *The Aharonov-Bohm effects: Variations on a Subtle Theme,* Physics Today: 38-3. *"Thus far such experiments, crucial as they are to the characterization of the AB effects, have remained out of reach. Nor has the pulsed version of the original (Type I) electric AB effect have been performed."*

194   W. Shockley and R.P. James, *"Try Simplest Cases" Discovery of "Hidden Momentum" Forces on "Magnetic Currents,"* Phys. Rev. Letters **18**, 876 (1967).

ceived a great deal of attention among scientists. The problem was considered unsolved for almost thirty years, until Comay published an explanation for this paradox in 1995.[195] The explanation is based on fundamental laws of special relativity.

On this occasion Comay had no problem publishing his explanation in a mainstream journal. This is probably because he made no attempt at discrediting a dogmatically held theory.

Problems with the Standard Model

In the early 2000s Comay was studying the Klein-Gordon equation, and he found it to be inconsistent. The meaning of this is that not only is there no particle that conforms to it, but the equation itself is unable to describe any elementary particles.[196,197]

Most notably, he showed that the way Yukawa had used this equation could not have described a massive particle[198] as Yukawa thought. Dirac also argued that the equation was incorrect.[199]

Although it is commonly accepted today that no particle that conforms to the Klein-Gordon equation exists, there are certain equa-

---

195 E. Comay, *Exposing "hidden momentum,"* American Journal of Physics, 1028–1034 (1996).

196 E. Comay, *Further Difficulties with the Klein-Gordon Equation,* Apeiron **12**, No 1, 26 (2005).

197 E. Comay, *The Significance of Density in the Structure of Quantum Theories,* Apeiron **14**, No 2, 50 (2007).

198 E. Comay, *The Yukawa Lagrangian Density is Inconsistent with the Hamiltonian,* Apeiron **14**, No 1, 1 (2007).

199 P.A.M. Dirac, *Mathematical Foundations of Quantum Theory,* Ed. A.R. Marlow (Academic, New York, 1978). p. 3,4.

tions "of the same family" that describe other elementary particles, such as the Higgs boson. Comay also showed that the Higgs boson equations suffered from the same inconsistency and were unable to describe an elementary particle.[200] He later showed that the equations that described the W particle, which was supposedly an elementary particle that carried an electric charge, were incommensurate with basic properties that ought to be found in every elementary particle with an electric charge.[201] By so doing, Comay demonstrated his objections to other parts of the Standard Model in addition to QCD.

To summarize, readers might have guessed by now that Comay's articles did not win him a lot of friends.

It is a fact that scientists who go against the mainstream school of thought are considered pariahs by notable journals in many fields of physics, not just particle physics.[202] Several scientific journals have been founded in order to counteract this trend.

---

200 E. Comay, *Physical Consequences of Mathematical Principles*, Progress in Physics, **4**, 91 (2009). See chapter 4.

201 E. Comay, *Quantum Constraints on a Charged Particle Structure*, Progress in Physics, **4**, 9, (2012).

202 Dmitri Rabounski, *Declaration of Academic Freedom*, (*Scientific Human Rights*), Progress in Physics, 1,57 (2006).

The Ivory Tower

The community of particle scientists is considered by many to be the elite among physicists. The theory held by particle scientists is deemed the most solid and accurate theory ever created by man.

I hope that all readers of this book now understand that there is something very wrong with a principal element in the predominant theory of particle physics. I assume that almost all readers would ask themselves whether it's possible that a normative scientific community would actually find itself in such a situation in the modern age. In this last chapter I will try and say something about the conditions that have led to this curious situation.

What *Is* Particle Physics?

The definition of particle physics requires certain clarifications. In the last century several major splits and mergers of different physical branches have taken place. For this reason, it is important to define the purview of particle physics.

The physical field closest to particle physics is called nuclear physics. Some universities have combined nuclear physics with particle physics.

Nuclear physics uses the properties of protons and neutrons (nucleons) to describe the characteristics of nuclei composed of several nucleons. These scientists are rather practical in nature. A nuclear physicist generally does not work on the internal structure of nucleons or on the laws that govern it.

The reason this is considered a practical approach is that nucleons are composite particles and the nuclear force is a residual and very complicated force. Therefore, it is treated on a phenomenological basis. As far as a nuclear physicist is concerned, there is little use in deciphering the internal forces inside nucleons, particularly because the Standard Model is so mathematically complex that it is impossible to apply it to the study of atomic nuclei.

Another branch close to particle physics is astrophysics. Astrophysics deals with the properties of astronomical objects. Astrophysicists use ideas taken from the Standard Model in order to try and guess how astronomical processes take place, and to try and describe hypothetical stars that are composed of materials that are different than what we presently know.

The Tower of Babel

As I mentioned in the beginning of this book, the founders of QCD created a new language, as they believed that the old physics was unable to describe the quantum states of particles belonging to the nucleon family. Freeman Dyson, one of the developers of quantum field theory, candidly said that the new physicists were using a language that he could not comprehend. Dyson did not feel the need to understand that language, as by that time he had turned to other fields of physics.

This new language comes in addition to the old physics. Therefore, particle physics must be familiar with other branches of physics, such as the spectroscopy of atomic electrons.

The problem is that particle physicists are absolutely convinced that the forces with which they are dealing, the strong forces, are unlike anything else found in nature. Therefore, so I assume, they simply neglect the other branches of physics. Perhaps this neglect also stems from the tremendous efforts they invest in studying and understanding the Standard Model in general and QCD in particular.

In the course of this book I have presented an example where famous physicists made a few basic mistakes as a result of this ignorance. The example is that particle physicists are unfamiliar with the theory of Wigner and Racah, and therefore do not know that nature does not contain bound particles with three fermions or more that are found in a pure s-wave state. Fritzsch, one of the inventors of QCD, made that mistake in an article in which he reviewed the history of QCD.[203] Another example is Yukawa's theory, which contradicts the basic principle of parity conservation.

Technology

The various branches of physics, other than astrophysics and modern particle physics, have made a decisive contribution to the technological advancement of mankind. Therefore, physicists from other branches, such as solid state physics, medical physics, or nuclear physics, habitually consider the significance of human necessity in the course of their work. Almost every day these physicists ask themselves how their research may, should it suc-

---

203 Harald Fritzsch, The history of QCD, CERN courier, September 27, 2012. *"The Ω^- is a bound state of three strange quarks. Since this is the ground state, the space wave-function should be symmetrical."* (Here "symmetrical" means ground state s-wave.)

ceed, bring about new insight that may later allow the birth of new technology.

The affinity between physicists and technology imposes a burden on the scientists. They must do more than amuse themselves with abstract ideas, but also see themselves as messengers whose role is to provide new insights that may later benefit all mankind.

Particle physics, however, does not share this burden. It is entirely unclear how the understanding of forces operating inside protons could ever bring about the birth of any future technology. As early as the 1960s, Haim Harari, a particle physicist at the Weizmann Institute, argued in his lectures that particle physics would no more contribute to the development of new technologies than would the philharmonic orchestra. So far, he appears to be right.

Isolation

It eventually came to pass that the vast majority of physicists had absolutely no idea what particle physicists were saying, and there is no organization at present that expects anything practical to be derived from research in the field. Particle physicists rarely interact with physicists from other branches, except for astrophysicists, who make use of ideas related to the Standard Model. A particle physicist does not contribute to any enterprise or productive company. A particle physicist can remain so only in academia.

Under these circumstances, particle physicists are able to pursue their research almost without having to face any external pressure. Are these conditions ideal for scientists? Probably not.

Non-conservation of Knowledge

The dozens of Standard Model contradictions presented in this book are almost entirely absent from textbooks. These contradictions were published in scientific articles written by mainstream scientists who almost always believe in the veracity of QCD.

Therefore, it is only natural that modern particle scientists, who are unaware of those contradictions, will say that each and every experiment with modern particle accelerators is fully consistent with Standard Model predictions. Others may mention a certain phenomenon or two that fail to conform to the model. One professor,[204] for example, argues that the only phenomenon that fails to fit with the model is the strong CP problem.

And what of the Nobel Prize laureate Sheldon Glashow, who argued that the results of a certain experiment were a thorn in QCD's side? And of another Nobel Prize laureate, Frank Wilczek, who admitted that QCD completely contradicted fundamental nuclear phenomena? Apparently, in the world of particle physics, this is business as usual.

Gag Order

In 2007 John Ellis, one of the leading scientists at CERN, was asked what would happen if the Higgs boson weren't discovered in the next experiment. The Higgs boson must exist according to

---

204 profmattstrassler.com/articles-and-posts/particle-physics-
basics/c-p-t-and-their-combinations/.

the Standard Model. Ellis responded by saying that if the Higgs particle were not detected, then "we theorists have been talking rubbish for the last thirty-five years."

Ellis has published a very large number of articles that relate to the Standard Model. He reacted spontaneously, which shows how confident he is of the veracity his theory. Ellis did not reject the possibility that Standard Model theoreticians have been speaking rubbish for so many years.

On the other hand, even without my asking Ellis, I am convinced that he completely rejects the possibility that special relativity is wrong. Here I present special relativity as an example of a physical theory that all physicists believe to be true.

However, in spite of the many documented Standard Model experimental failures, an examination of articles published over the last few decades in notable and leading physics journals will not turn up a single one that objects to the Standard Model. One article that cast doubts on special relativity *did* get published in 2011, and it caused a substantial outcry.[205] It was later found that the data on which the article was based were measured by a faulty device.

By the way, articles that contain alternative explanations are published in many cases in an archive called "arxiv," which allows well-known physicists to publish their works even if they aren't accepted by official journals.

---

205 OPERA experiment reports anomaly in flight time of neutrinos from CERN to Gran Sasso, CERN, September 23, 2011.

Violence

Academia should constitute a pleasant environment. However, those who dispute the Standard Model fall victim to repugnant aggression at the hands of those defending the theory. A suitable environment for academic debate is not afforded to those who dare question such cherished beliefs.

I put a lot of thought into whether I should quote such instances of aggression, or whether I should leave them out of this book. I chose the second option. Readers who are fans of tabloid newspapers can simply Google phrases such as "fuck you" and "idiot" added after my last name. These superlatives were given to me by distinguished professors or graduate students (or so it appears, considering the breadth of knowledge they demonstrated). The physical arguments accompanying this verbal abuse were not particularly convincing.

No wonder, then, that almost no one would dare dispute the Standard Model.

They Can't All Be Wrong

I assume that petulant and aggressive academics are actually unfamiliar with the many contradictions to their theory. Almost none of these contradictions can be found or have never been found in textbooks. Such academics are thoroughly brainwashed by authors who consider the Standard Model to be the greatest scientific achievement of all time. The best warriors are those who believe in their cause. And there are quite a few who fight to defend the Standard Model.

A physicist who was formerly the head of the particle physics department at Tel Aviv University received ten lines of text showing a theoretical mistake in one of the premises on which the Higgs boson equation, among others, is founded. Ten lines! What can be easier than refuting an argument that spans less than a page if the Standard Model is indeed so devoid of contradictions?

Several days later the "refutation" was finally given by the esteemed professor: "If what I read here is true, then everybody's wrong. And the chances that everybody's wrong are one in a million!" In my experience, the department of mathematics would have rejected such "proof" outright.

In a scientific community where no one takes responsibility, and the main opposing argument is that the "herd" is always right, it is no wonder that an intelligent community of scientists has found itself in one of the strangest situations in the history of science.

What's Next?

One way to give scientists a ladder they can use to climb back down to earth and eventually undermine the universal support of QCD is by conducting a crucial experiment, each result of which would refute either the liquid drop model or QCD. In a previous chapter this was referred to as a "decisive experiment."

Would particle scientists rise to the challenge and agree to go through with the experiment, even though they have no interest in doing so?

If so, that perhaps will be the first stage in reuniting particle physics with the other branches of physics. The next stage will be to include the very long list of phenomena that contradict QCD in textbooks and teach them in universities. Only then will it be possible to finally pursue a free academic discussion on the theory that accurately describes nature's fundamental particles.

Who knows, perhaps such alternative scientists will be able to fulfill their dream and see the entire scientific community confess to a mistake it made more than forty years ago.

> "My dream is to confirm the magnetic monopole model of hadron by observing directly the monopoles of opposite sign fuse to form a meson."[206] –Tetsuo Sawada, 2003

---

206 Tetsuo Sawada, *Is the nuclear force short range?* arXiv:nucl-th/0307023, (2003).

EPILOGUE

The Cave

In this chapter I present an analogue to the cave allegory found earlier in this book.

In the 1960s it was discovered that protons and neutrons contained smaller particles known as quarks. But things didn't work out, and the way quarks interacted with each other was unlike anything scientists knew about other natural structures such as atoms. Scientists then made two assumptions:

1. The only massive particles inside protons and neutrons were three quarks.

2. All quarks found in the lighter particles are s-wave states situated at its lowest energy level. (The fact that the three quarks occasionally had parallel spins contradicted a natural law known as the Pauli principle.)

The discovery troubled scientists at the time. But in 1972 a new and revolutionary idea was published in scientific literature. A group of scientists formulated a new theory based on entirely new laws hitherto unknown to science. The theory was called quantum chromodynamics, or QCD.

Asymptotic Freedom

Despite the fact that the theory seemed fantastic and was unlike anything else known in nature, one of the scientists who developed the theory was a leader in the field, and therefore researchers decided to give it a chance and see if it fit with other data obtained by particle accelerators.

One of the theoretical problems of electromagnetic forces is known as the Landau pole. The problem is caused by the fact that the force operating between two electric charges grows as the charges draw close to one another, and, according to the equations, could become infinite as they reach a negligible distance from one another. However, an examination of QCD's equations revealed that the force operating between quarks grew stronger as the quarks drew away from each other—the exact opposite of any other force known in nature! Therefore, the theoretical problem of the Landau pole did not affect the forces operating between quarks.

This property of QCD is known in literature as asymptotic freedom, and it is to this day considered one of the most well-founded proofs of QCD's validity. In 2004 the three discoverers of asymptotic freedom were awarded a Nobel Prize for finding this QCD behavior.

It wasn't long before the new theory triumphed once more. Scientists who studied QCD predicted that an analog to the bremsstrahlung phenomenon found in electromagnetic forces existed in strong forces and would result in the formation of three jets. In 1978, a three jet event was discovered in the process of electron-positron collision. This phenomenon, known as the three jets event, is considered proof of QCD's validity, despite the fact that another known force, the electromagnetic force, behaves in a similar way.

Contradictions

Even upon the invention of QCD, scientists knew about the great similarity between the force operating between nucleons and that operating between noble gas atoms. This similarity remains unexplained to this day. Leading scientists in the field have said that the similarity appears to contradict QCD.

In 1974 it was discovered that quarks carry only about one-half of the nucleon's mass. This discovery astounded scientists since according to QCD there were no massive particles inside nucleons apart from the three quarks. But scientists recovered and claimed that the missing mass was carried by gluons.

According to QCD the interaction between high-energy polarized protons should be identical regardless of whether their spins are parallel or opposite. But in an experiment conducted in 1977 by the team of Michigan Spin Physics Center, it turned out that the interaction between protons with parallel spins was much stronger than that occurring between protons with opposite spins. The Michigan researchers were then asked to see if this also happened in higher-energy protons, which turned out to be the case, which means that the contradiction to QCD remained. But QCD proponents argued that this finding should not be regarded as a contradiction to QCD as it may be explained in the future.

Scientists employed QCD to predict that quarks of nucleons inside the nuclei of large atoms would have a smaller volume than that of nucleons situated inside very small atoms. In an experiment conducted by the EMC in 1983, however, it turned out that nucleons actually had greater volumes in nuclei of large atoms. This discovery contradicted the hypotheses published before it was made and it remains unexplained to this day.

In 1987 the EMC conducted an experiment with polarized muon beams and discovered an unexplained effect known as the proton spin crisis. A possible explanation considered by scientists was that the quarks inside the proton were never characterized by a pure s-wave. This discovery also remains unexplained to this day.

In the 1990s it was found that the cross section describing the collision of electrons and positrons with protons was larger than expected. This discovery provided an indication for the existence of massive particles inside the proton. In the early 2000s scientists found a rise in the cross section graph of elastic proton-proton collisions. This discovery provided irrefutable evidence of the existence of massive particles inside the proton. It remains unexplained to this day, as according to QCD there are no massive particles inside the proton other than quarks.

Statistical measurements of the distribution of quarks inside the proton showed that quarks tend to be found more at the center of the proton rather than in its periphery, similar to the distribution of electrons inside atoms. This discovery seemed to contradict the concept of asymptotic freedom, as it appeared that similar to the electromagnetic force the strong force intensified the smaller the distance between quarks became. In a cross section measurement of proton-proton collisions it was found that the force operating between the quarks of one proton and the quarks of another intensified as the distance between the protons decreased. This too appeared to contradict QCD, which predicts the existence of asymptotic freedom, according to which the force operating between quarks ought to weaken as the distance between them decreases.

QCD experts also predicted the existence of exotic particles, such as glueballs, dibaryons, pentaquarks, and strange quark matter. However, to this day no such exotic particles have been found despite extensive search efforts.

Several years before the quark discovery, physicists found that energetic photons interact strongly with nucleons and that the intensity of the photon-proton interaction is very similar to that of photon-neutron. Since physicists assumed that pure photons only interact with electric charge, they adopted the Vector Meson Dominance (VMD) theory, which claims that a photon is a superposition of a pure electromagnetic photon and vector mesons. The VMD explanation was taken seriously for few decades, despite the fact that it is inconsistent with special relativity. Later the VMD explanation was abandoned and removed from textbooks. The photon-nucleon interaction was removed as well, probably because it cannot be explained by QCD.

Summary

So what did we see here? First, the two assumptions that led to the sixties crisis were:

1. The entire mass of the proton is carried by quarks.

2. All quarks are characterized by a pure ground state s-wave.

These two assumptions are now very suspect. Because of these assumptions it was believed that the old physics could not describe the internal structure of the proton, and therefore the QCD was needed.

Furthermore, all of the discoveries made throughout the years have showed that the properties of nucleons are very similar to those of ordinary atoms. It seems to contradict QCD, which assumes that the properties of these particles are very different from those of known forces.

What Do the Scientists Say?

Scientists continue to regard QCD as the only possible theory. When confronted with seemingly contradictory data, QCD scientists say that QCD calculations are so complex that even the fastest computers will likely never be able to accurately utilize them.

Moreover, say the scientists, lattice QCD has been very successful in explaining the masses of baryons and mesons. Therefore, they believe, QCD must be true. Other models that do not assume the veracity of QCD also arrive at accurate results when used to predict the masses of mesons and baryons, such as the Gell-Mann-Okubo mass formula.

Mainstream scientists today are all very pleased with QCD theory. Some argue that even if no reasonable explanation is found in the future for all the phenomena mentioned thus far, they will nevertheless refuse to evaluate any other theory, and we will be forced to content ourselves with the knowledge that QCD's equations are so complex that we'll never be able to calculate the properties derived from them with sufficient accuracy.

Moreover, there are thousands of authors who consider QCD to be a principal element of the most accurate theory ever invented by mankind, in any field. Nothing less than that!

And that, in a nutshell, is QCD.

INDEX

A

B

C

D

E

F

G

H

K

L

M

N

P

Q

R

S

www.ingramcontent.com/pod-product-compliance
Lightning Source LLC
Chambersburg PA
CBHW060543200326
41521CB00007B/461